An articulate and novel take on energy security both as an analytical concept and as a lens through which to examine the complex dynamics and values affecting energy trends in the United States. Important reading for anybody seriously interested in the future of energy security in the United States, especially those striving to make America's energy system more environmentally friendly, renewable, and truly secure.

Benjamin K. Sovacool, University of Sussex, UK

A major contribution to the energy security literature. Bernell and Simon are rare *boundaryspanners* who have been operating at the intersection between the worlds of practice and the academe (Bernell), and between the worlds of science and engineering and social science policy studies (Simon) for over two decades. Energy security policy necessitates a broad range of knowledge and experience, and this book captures the insights generated from this truly collaborative work reflecting a remarkable scope of familiarity with the political, economic and scientific dynamics of energy security.

Nicholas Lovrich, Washington State University, USA

Bernell and Simon have crafted an engaging and thoughtful examination of energy security that both captures the twentieth-century geopolitical roots of the topic and ties them to the emergent needs of the globalized twenty-first-century. It's a brave volume that isn't afraid to tackle incongruence and contradiction with candor and insight, in one of the central political conflicts of our era.

Adam L. Reed, University of Colorado-Boulder, USA

THE ENERGY SECURITY DILEMMA

This book analyzes the energy security of the United States—its ability to obtain reliable, affordable, and sufficient supplies of energy, while meeting the goals of achieving environmental sustainability and protecting national security. The economic and national security of the United States is largely dependent upon fossil fuels, especially oil. Without significant changes to current practices and patterns of energy production and use, the domestic and global impacts—security, economic, and environmental—are expected to become worse over the coming decades. Growing US and global energy demands need to be met and the anticipated impacts of climate change must be avoided—all at an affordable price, while avoiding conflict with other nations that have similar goals.

Bernell and Simon examine the current and prospective landscape of American energy policy, from tax incentives and mandates at the federal and state level to promote wind and solar power, to support for fracking in the oil and natural gas industries, to foreign policies designed to ensure that markets and cooperative agreements—not armies, navies and rival governments—control the supply and price of energy resources. They look at the variety of energy related challenges facing the United States and argue that public policies designed to enhance energy security have at the same time produced greater insecurity in terms of fostering rising (and potentially unmet) energy needs, national security threats, economic vulnerability, and environmental dangers.

David Bernell is Associate Professor of Political Science at Oregon State University, USA.

Christopher A. Simon is Professor of Political Science at the University of Utah, USA.

THE ENERGY SECURITY DILEMMA

US policy and practice

David Bernell and Christopher A. Simon

Routledge
Taylor & Francis Group

NEW YORK AND LONDON

First published
by Routledge
711 Third Avenue, New York, NY 10017

and by Routledge
2 Park Square, Milton Park, Abingdon, Oxon OX14 4RN

Routledge is an imprint of the Taylor & Francis Group, an informa business

© 2016 Taylor & Francis

The right of David Bernell and Christopher A. Simon to be identified as authors of this work has been asserted by them in accordance with sections 77 and 78 of the Copyright, Designs and Patents Act 1988.

Library of Congress Cataloging in Publication Data
Names: Bernell, David, author. | Simon, Christopher A., 1968- author.
   Title: The energy security dilemma : US policy and practice / David Bernell, Christopher A. Simon.
   Description: New York, NY : Routledge, 2016.
   Identifiers: LCCN 2015042151| ISBN 9780415890540 (hardback) | ISBN 9780415890557 (pbk.) | ISBN 9780203817797 (ebook)
   Subjects: LCSH: Energy security–United States. | Energy policy–United States.
   Classification: LCC HD9502.U52 B477 2016 | DDC 333.790973–dc23
LC record available at http://lccn.loc.gov/2015042151

ISBN: 9780415890540 (hbk)
ISBN: 9780415890557 (pbk)
ISBN: 9780203817797 (ebk)

Typeset in Bembo
by Taylor & Francis Books

Printed and bound in the United States of America by Publishers Graphics, LLC on sustainably sourced paper.

CONTENTS

LIST OF ILLUSTRATIONS

Figures

Tables

Boxes

PREFACE

Picture the authors of this book, one in Corvallis, Oregon and the other in Salt Lake City, Utah, sitting at tables in coffee shops where they like to write. This is a common scenario for them, and for many of us. On one day one of the authors, David, turned to a woman sitting nearby him and said "Excuse me, could I please use the electrical outlet to plug in my laptop? It's running out of power." On that particular day, the demand for electricity had exceeded the supply of electrical sockets. "The irony," he said, "is that I'm writing a book about energy security, but I don't have enough energy security to write the book!"

The truth, in this case the truth of the societal importance of energy, gets revealed in many, often mundane ways, such as in finding a nearby outlet, and irony is a particularly powerful force in revealing truth. The central idea put forth in this book is that in the United States, our seemingly insatiable demand for energy, at all times and in just about all places, with minimal cost and adverse impact, presents a major public policy dilemma. We cannot have everything we want, and difficult tradeoffs cannot be avoided despite continuing breakthroughs in technology and ever more alternative energy sources being developed. There is always a tradeoff and a price to pay, whether these things are acknowledged or ignored, paid in full, in part, or pushed off onto others to pay. In the case of the coffee shop electricity supply problem, the tradeoff was simple—a little "demand management" solved the problem, whereby the woman sitting near David removed her power cord from the socket and allowed him to use it for a while to charge up his computer.

On a much larger scale, however, in thinking about energy security as a challenge with which we grapple as a nation, the tradeoffs and costs are much more significant, and the problem itself eludes a simple solution. The contradictions are all around us. On the one hand, the world is literally awash in energy of all types.

The end of fossil energy may or may not be on the near horizon, but for the time being, massive consumption goes forth on a global scale. As of 2015, the world is regularly consuming, on an annual basis, almost 32 billion barrels of oil (which translates into the more well-known figure of 90 million barrels of oil per day), more than 8 billion short tons of coal, and over 120 trillion cubic feet of natural gas, although the latter statistic is, at the very best, an informed guess given the unknown quantities of natural gas that are simply "flared off" at petroleum well facilities. The world procures almost 60 thousand tons of uranium ore per year used to produce radioactive fuel rods for nuclear power plants and, incidentally, nuclear weapons. And, of course, electricity continues to flow every day. Over 20 trillion kilowatt hours of electricity were consumed worldwide in 2014, and that figure will continue to grow as the developing world—particularly the BRIC countries (Brazil, Russia, India and China)—gain greater access to the benefits of a middle class lifestyle.[1]

At the same time, the United States seems to be in the midst of a veritable energy revolution. In spite of the (perhaps temporary) drop in oil prices beginning in 2014, great quantities of energy pour forth from wind farms in Texas, oil fields in North Dakota (though this has been greatly impacted by the drop in the price of oil), solar panels in California, and natural gas wells in Pennsylvania. The outward benefits of the energy development and production involved can be seen all around. Oil production is booming—the United States was the largest petroleum-producing nation in the world in 2014, passing Saudi Arabia and Russia. Oil imports have dropped to levels not seen in decades, while prices have dropped due to high global production. Renewable energy use is at an all-time high; greenhouse gas emissions from power plants drop as cheap natural gas increasingly replaces coal; and billions of dollars in the energy sector are being made, saved, and invested. Such good fortune provides a welcome break in the United States from longstanding concerns about energy security, but it also contributes to a certain complacency setting in, as we come to believe that new technology has brought the country a significant increase in petroleum and natural gas, and along with it, greater security.

When one looks, however, beyond the US domestic oil and gas fields at the continued political chipping-away at support for renewables in Congress and several states, at climate change legislation, at a European Union often distracted from its own long-term clean energy agenda due to Russia's actions in the Ukraine, at the political turmoil, especially the rising power of the Islamic State in Iraq and Syria, and at China making clear its claims to sub-ocean petroleum and gas fields in the South China Sea, the larger context comes more clearly into focus. Any contemporary smugness born of short-term technological breakthroughs and good fortune is likely to be short-lived and is most certainly unwarranted. The world may be awash in energy, but the world likewise is a dangerous place with multiple threats to these supplies and their continued patterns of use.

So despite the sense that in the energy arena some things are better than before, there is a palpable sense of uneasiness about the longer term future, a growing awareness that such enormous consumption of energy resources cannot continue unabated, that we must reduce our nation's dependence on fossil energy. In 2013, a Gallup poll found that 51% of individuals polled in the United States favored environmental protection over conventional energy development in a direct tradeoff. Gallup also found that Americans favor conservation of energy over energy production in a comparable tradeoff framing of choice.[2] State and federal government public policy commitments to green energy and efficiency reflect this support and have grown tremendously in recent years—witness the federal tax benefits and the 38 US states with either enforceable renewable portfolio standards or softer "goals." Still, these attitudes and policies have not yet translated into the large scale behavioral and policy change needed to part ways with fossil energy dependence and a vulnerable electricity grid. This set of circumstances confronts us with a choice—namely, we can either make the changes in a pro-active manner, or the threat of war and terrorism, human-caused accident, or natural disaster will force us to reconsider our actions in relation to our stated priorities of conservation and support for clean energy. In this regard, perhaps Winston Churchill captured one aspect of our national character better than anyone else, offering both hope and dread for the future, when he observed that, "The Americans can always be counted on to do the right thing, after they have exhausted all other possibilities." Considering all these concerns, it is not surprising that we cannot seem to stop arguing in this country about energy policy and energy security. Nor should we. In spite of all the benefits the United States is experiencing at this juncture in our history, the unavoidable tradeoffs and societal and environmental costs associated with each of these benefits, along with emerging threats, are very real, and potentially very large.

Of growing importance is the opportunity, or perhaps necessity, to revise the way we think about choices and their actual costs. Price is the link between supply and demand, but what of marginal social and environmental costs? While the common response is that such costs are difficult to measure, markets are able to divine the seemingly inconceivable on a near minute-by-minute basis. For example, how often have you read about the market getting "concerned" or "worried" about the words of a world leader, an event in some far-flung corner of the planet? Somehow, a "cost" can be attached to these emotions, these concerns. So why not for social or environmental costs? Outside of the market, of course, these costs are attached to cases involving corporate-caused disasters. Witness the $18 billion settlement in the BP Deepwater Horizon oil spill in the Gulf of Mexico.

An *ex ante* policy approach would require rethinking the way land, water, and natural resources are accessed and the royalty structure paid by energy developing firms, be they fossil energy, uranium, or other forms of mining or renewable energy firms operating on public lands. In fact, it might suggest a rethinking of longstanding views of sub-surface and surface mineral rights and property rights, more generally. Rather than wait for a disaster, would it not be better to consider these costs upfront, along with the tradeoffs that may be required, as they would likely help policymakers in their endeavors, allow individuals to make more cost-effective decisions about energy consumption, and give corporations a clearer sense of the costs of doing business (as either energy consumers or producers)? We believe that it is worth our time, effort, money and thinking to consider these difficult matters, and that is what led us to write this book. It is the very nature of these costs and tradeoffs, what we call an "energy security dilemma," that forms the subject of this book.

A moral dimension is of equal concern in the consideration of a long-term change around energy development. Energy security is about regular supply and wise use of resources, but energy security is much more than that. Energy security is also peace of mind, knowing that our stated values match our actions, and that our actions bring benefits to us and others, not harm. Some of the violence in the Middle East today is being financed by petroleum sales made by ISIS to feed its militants, to pay for the "beans and bullets" needed for their dream of an Islamic Caliphate. Someone is purchasing that petroleum and many parties are benefiting from the commercial exchanges involved. Whether or not we, individually, purchase a barrel of oil that funds war, terrorism and violence, the market supply, demand and prices are influenced by this nefarious oil sale. Therefore, if we believe in diminishing the harmful impacts of our energy choices, in weakening the position and attractiveness of global violence, in environmental protection, in promoting energy security in all its many dimensions, it is imperative that we understand energy security and the most appropriate ways of promoting it.

We would like to thank our families and colleagues for their support in our multi-year adventure in collaboratively drafting this book manuscript. David is dedicating this work to his wife Steph, and his boys, Eli and Miles, who remind him on a daily basis how lucky he is, and why caring about the future is of the greatest importance. Chris is dedicating this work to his parents, Raffi G. and Susan M. Simon, who have always been there for him. They have great discussions about politics, economics, society, and morality on a daily basis. He always benefits from their wisdom and knowledge. We would also like to thank the School of Public Policy at Oregon State University and the Department of Political Science at the University of Utah, and our colleagues, for their support as we have marched through this very challenging topic. Finally, we would like to thank Nicholas P. Lovrich, Regents Professor Emeritus, Washington State University for reading multiple drafts of this manuscript, offering outstanding editorial suggestions and advice, and for acting as a form of "umpire" between us, helping us to discover that our thoughts and values were more closely aligned than we had previously imagined. Thank you, Nicholas.

Notes

1 Energy Information Agency, "International Energy Statistics," Washington, DC: U.S. Department of Energy, www.eia.gov/cfapps/ipdbproject/IEDIndex3.cfm?tid=1&pid=1&aid=2# (accessed January 29, 2015); and World Nuclear Association, "Uranium Markets," www.world-nuclear.org/info/nuclear-fuel-cycle/uranium-resources/uranium-markets/ (accessed January 29, 2015).

2 Gallup Organization, "Americans Pick Environment over Energy Development for the First Time in Three Years," Washington, DC: Gallup Organization, www.gallup.com/video/171008/americans-pick-environment-energy-development-first-time-three-years.aspx (accessed September 4, 2014); and "Americans Still Favor Energy Conservation over Production," Washington, DC: Gallup Organization, www.gallup.com/poll/168176/americans-favor-energy-conservation-production.aspx (accessed September 4, 2014).

1

ENERGY SECURITY AS A CONCEPT

If you have never read the Jack London short story, "To Build a Fire," we recommend it. If you have read it, then we suggest you read it again, but this time view it as a metaphor for energy security. It is a story about a human being standing face-to-face with the natural world. The man in the story travels with a dog through the cold Arctic wilderness. The man is warmly clothed in animal furs, but his clothing is not sufficient to protect him when he accidentally gets his feet wet. He attempts to build a fire, a fundamental source of thermal energy, to dry his footwear and warm himself, but fails for lack of care in the use of abundant fuel all around him. Failing to make plans appropriate to his environmental challenges he ends up freezing to death. The dog with which he travels is, in comparison, more self-sufficient; she trots off along a familiar trail toward the warmth and comfort of civilization. The man's arrogant belief that he commanded the natural world—that he was the master of himself and his surroundings—proved to be a seriously flawed illusion. In walking alone in the frigid Arctic weather, he assumed risks so great that one false move, one small accident, led to horrific outcomes. Without needed energy resources and without good planning and an awareness of the risks to his security, he perished in the wilderness.

It is our contention that the man in Jack London's story can be viewed as a metaphor for modern society. Civilized society assumes an enormous risk every day in which we collectively rise from our sleep in the belief that there will be ample resources to meet our needs in the form of energy for agricultural harvests, for cooking, for warmth and cooling, for transportation, and for the manufacture and provision of desired goods and services. There is clear evidence that the energy resources we demand now and will require in the future are at present available, but the regular supply of these energy resources requires careful thought and planning as we move into the future facing the dual challenges of a growing global population and climate change phenomena posing additional risks for many areas of the world. In the case of the man in the wilderness, he foolishly built his fire under a snow-covered branch. As the heat of the fire rose, the snow on the branch fell and put out the lone fire he was prepared to build. In the case of civilized society, we have put our faith in energy supply chains that extend half way around the globe, in an electrical grid that we continually seek to upgrade and make more secure after each "brown out" episode, and in finite sources of fossil energy that produce emissions detrimental to human health and potentially could lead to great planetary harm. In other words, we have been for a very long time arrogant and unjustifiably optimistic in our energy demands, ignoring the often delayed or hidden costs and tradeoffs inherent in these demands and the quest for energy security.

If we could transport ourselves back in time to ancient Mesopotamia, or even to 18th-century Europe or Asia, we would find human societies consciously accessing energy resources to enable increasingly civilized existence, in much the same way as modern society does but on a far smaller scale. As with all human societies, a division of labor arose that brought together human expertise, sources of labor (human and animal), and needed energy resources to make possible the well-orchestrated production of goods and services in population centers of ever-increasing scale. "Water masters" managed rivers and streams to flood croplands, but water also provided needed energy inputs—principally to turn gristmills for grinding grains into flour for bread, pasta, tortillas and noodles. Others secured lamp oils to provide light for city streets, and for individuals whose work required them to toil in the hours before dawn or past dusk.

In the 19th and 20th centuries, energy needs grew as mechanization (and later computerization) changed the nature of work and the division of labor. The large-scale commercialization of coal and petroleum extraction in the 19th and 20th centuries made possible the growth of modern cities and the development and expansion of motorized transportation and electrification. And with these changes, we collectively assumed yet greater risk. No longer relying solely on the time or expertise of women and men, or on animal power, we became more reliant on an ever-widening variety of machines, each requiring some form of energy input, to aid in the provision of food, warmth, shelter, manufacturing, transportation and entertainment.

For humankind, access and direct use of abundant energy resources has played and continues to play a central role in social, political and economic evolution from the primitive to the modern era. *Energy security*, therefore, involves a great deal more than protecting a resource. It is about maintaining all that we have gained—namely, technology, transportation, manufacturing, electricity, buildings, information technology, medical advances—long into the future. All of this abundance-enabling societal infrastructure and more is possible because we have been able to rely upon vast supplies of energy. It follows that energy security can and must play a central role in all that we will likely gain in the future. Without abundant supplies of energy and reliable access to these supplies, human progress would likely slow down, come to a halt, or be reversed outright.

Energy security, however, entails more than relative abundance and ready access for a nation or the global community collectively considered. The purview of energy security is larger than that. In particular, with regard to the United States—the focus of this analysis—it involves a range of concerns comprising energy type and quality, affordability, cleanliness vis-à-vis environmental degradation, diversity of sources, and the reliability of the energy supply chain and infrastructure. In other words, as an objective of pursuing national interests both at home and abroad, we posit that *energy security refers to the ability to obtain abundant, reliable, affordable, diversified supplies of energy (in terms of both fuel types and their geographic sources), while also meeting the goals of achieving environmental sustainability and enhancing energy independence.* (Energy independence is itself a controversial term, and means different things to different people—usually something in between total energy self-sufficiency and the diminished role of foreign-supplied oil as a strategic commodity that has outsized impacts on the American economy and national security policy.) This is a tall order, especially as "the rise of the rest"—a phrase that Fareed Zakaria has popularized to refer to the rapid growth of China, India and other developing countries around the world—leads to greater economic activity, a growing global middle class, increased production and mass consumption of manufactured goods, and of course, greater demands upon the world's energy resources to fuel this remarkable and historically unprecedented economic growth.[1]

In this context, the world's largest energy user—the United States—can expect to have to devote considerably greater time, more attention and additional national resources to the problem of enhancing its energy security. At present the economic and national security of the United States is largely dependent upon fossil fuels, especially oil, and both the global demand for and competition over these resources is growing. Without significant changes to current practices and patterns of energy production and use, and the policies that govern such production and use, the domestic and global impacts—security, economic, and environmental—are expected to become particularly acute over the coming decades. Growing US and global energy demands need to be met, and the anticipated impacts of global warming must be avoided, all at an affordable price while avoiding conflict with other nations that have similar goals. The various ways to do this in terms of both conservation and enhancement of new and existing sources are hardly a mystery, but they all tend to be remarkably difficult to achieve in practice, both domestically and globally. Reducing demand is difficult in the context of the progressive globalization of trade and development; increasing and diversifying energy sources is occurring at a pace barely keeping pace with new buyers; and making the acquisition and use of any energy supplies (current and in development) a cleaner venture is an ongoing challenge throughout the world.

It is this understanding that informs the central argument of this book, which is that the effort to achieve energy security contains its own contradictions and obstacles. This is what we call the *energy security dilemma*, a circumstance in which the pursuit of such security is understood to involve goals and tradeoffs that can be inherently contradictory, as well as costly and inconvenient. Rather than focusing primarily on geopolitics, our aim is to examine this dynamic primarily in a domestic context, focusing on the dilemma as it plays out in the United States. In doing so, we look at things through the lens of public policy, which not only frames the context in which energy markets operate, but which is also both a response and a contributor to this dilemma. To that end, our analysis seeks to capture the wide array of concerns and goals bound up in the concept of energy security, which contains political, economic, technological and social dimensions.

The global economy and national societies have evolved quickly into energy-dependent entities in the modern era. Societal evolution in this regard is accelerating to a high level of sophistication of production and consumption networks of vast scope, with some of the greatest changes witnessed in the last half century. A rapidly moving global society often demands energy resources that produce what we want and need on demand, regardless of the short- or long-term risks involved. Given the nature of our ongoing and developing demands, energy access must be immediate and energy quality must be of an expected high level, in particular to properly power the computers, servers and information technologies that increasingly drive the global economy in nearly every corner of the planet.

On top of this, the affordability of energy supplies is essential, such that prices are not only within the means of those who want them, but that they are also fairly steady, without great volatility. High fluctuation in energy prices might lead to economic hardship, high rates of unemployment and even social upheaval, and in some countries it might even threaten regime stability.[2] Petroleum, natural gas, and coal are the three forms of hydrocarbon fuel that are today in high demand throughout the world because they meet our immediate energy needs in terms of abundance, access and affordability (though they are subject to much price volatility). As a consequence of these considerations, we Americans and others with whom we share the planet all seek greater access to such resources in spite of their toxic outputs, and in spite of the fact that establishing a continuous and secure supply of these needed resources creates new dilemmas for both the global and national environments and for global security.

At the same time, there are countervailing social developments that are both cause and consequence of our energy system, and these developments can have a substantial impact on resources. For example, beginning in the 1970s, the environmental movement brought with it widespread questioning of the long-dominant fossil energy paradigm. Built on growing evidence of the pollutant effect of fossil energy use and its adverse impact on human health and ecological conditions, as well as a generational value shift, this global social movement has led to long-standing public policy advocacy coalitions that support environmental regulation, the promotion of clean renewable sources of energy to power the post-industrial world, and the achievement of a more just and equitable political, economic, and social existence on a global scale.[3]

The market plays a critical role in both driving and responding to these diverse elements of energy security. The market for energy, however, is a highly complex global construct characterized by a multiplicity of competing interests and values. For example, petroleum, coal and natural gas, in particular, are treated as freely traded private goods on the global energy market. However, in fact these commodities are also goods belonging to the public, in that they offer significant social benefits. Contemporary politics, economics and social systems remain largely wedded in practice to a private market that provides a relatively narrow range of energy alternatives for direct or indirect use in the provision of the vast majority of basic goods and services associated with a modern economy. Moreover, the global energy market consists of an enormous and elaborate supply chain in the form of oil tankers, railroad lines, pipelines, and transmission grids that crisscross international borders. On top of this, there is a domestic network of pipelines, deepwater ports, natural gas offloading sites, railroads transporting coal, electricity power plants and transmission grids, all of which interact to deliver energy that meets our many demands involving source abundance, ready access, reliability in delivery and affordability vis-à-vis prevailing levels of income.

Alongside societal and economic drivers, technology is also an important element of how the energy sector evolves. We look to technology to help ensure energy security, as technology can lead to greater *abundance* (deepwater drilling, fracking, horizontal drilling, and new technologies to use renewable resources such as solar, wind, geothermal and biomass), *reliability* (distributed generation, diversity of fuel sources, modernization of the electric power grid), *efficiency* (smart grid, electric cars, more fuel efficient cars, fracking and directional drilling), *cleanliness* (renewables, carbon capture and storage) and *diversity* (domestically produced oil and natural gas, renewables). Because of their promise and potential to address societal, economic and environmental challenges, technological improvements—such as "clean coal" experiments with carbon sequestration at the site of power generation funded by both public and private sources—are increasingly sought as answers to the growing liabilities associated with the continued large-scale use of fossil fuels.

Lastly, there is a significant political dimension to energy security. As Harold Laswell so aptly described it, politics is the process of deciding "who gets what, when and how"—and with respect to energy, understanding this political element is both vitally important and challenging in the extreme. As society demands an abundant supply of reliable, modestly priced energy that does as little environmental damage as possible, we turn not only to the market and technology to meet these demands, but also to governments that will, through public policy, shape the conditions under which global and national markets operate to meet our growing demands, especially when expectations are not being met by the market-based incentives at play.

Some politicians and respected scholars challenge the legitimacy of this expectation, claiming that markets alone should dictate the supply and price of a good or service—be it energy or anything else—and that government policy tends to distort markets in inescapably unfavorable and unproductive ways. In his classic account, *The Road to Serfdom*, Friedrich Hayek details the dangers associated with large and powerful government seeking to protect citizen welfare.[4] Hayek argued that governmental action to this end actually has the opposite effect by denigrating individual freedom, which enjoys full bloom in the free market setting. In his classic treatise, *Capitalism and Freedom*, Milton Friedman parallels Hayek's line of argument as he outlines his arguments for a largely unfettered capitalism, one in which government regulation and market inventions are wisely kept to an absolute minimum.[5]

The aforementioned claim is challenged by arguments and evidence demonstrating the historical beneficial role of government in shaping markets, and the desirability of such government intervention in the economy to address the acknowledged problems of market failures, inequity, and the frequent adverse social impacts arising from the operation of a free market. John Maynard Keynes, whose work *General Theory of Employment, Interest and Money* was published in 1936, is broadly taken as the modern intellectual father of this type of activist approach to government's role in the economy. Beyond the economic arguments advanced by noteworthy economists, it is widely understood by scholars and public policymakers alike that governments—particularly in democratic societies—tend to represent the values of their citizens. In other words, governmental policy is a principal method of recognizing and reinforcing those preferred outcomes that are most important to citizens, among those being energy security.

For example, experience teaches us that in the United States when energy prices are low and/or people experience little fluctuation in reliable access, Americans are unlikely to become part of the political dynamic of energy policy. Conversely, when the opposite is true and there is a sharp rise in cost and/or shortages emerge in critical periods, energy becomes a central feature of domestic politics and can serve to delegitimize a system in the eyes of citizen-consumers, as politicians are called upon to "do something," even if the solutions only exacerbate the problem.[6] The oil shocks of the 1970s greatly affected consumer perspectives on energy as a policy priority, and had a major impact on public policy formation.[7] When petroleum supplies were tight and price spikes occurred, the United States imposed price controls and developed the Strategic Petroleum Reserve (SPR), and at various times since its creation US presidents have released petroleum reserves from the SPR in order to affect oil availability for refineries and prices at the pump. Politics have also led to a noteworthy change in the US energy portfolio, in which the country's leaders have consciously sought to move away from energy sources that may pose a risk in terms of either steady supply or environmental degradation. The American effort to promote the use of ethanol and other biofuels, beginning with the 1973 oil crisis and extending into the present day with the Renewable Fuels Standard, has been meant to address both of these concerns with respect to petroleum and related liquid fuels.

Sometimes energy markets themselves have proven to be a threat to energy supply, prompting public policy intervention. Perhaps the most egregious example of energy market failure in the US was witnessed in the wake of the deregulation of the California electricity market in the late 1990s, occasioning the rapid rise and precipitous collapse of Enron in 2001. Enron was one of the major energy traders in the California market debacle, and was active in energy markets nationwide. Enron first played a substantial role in policy changes leading up to the deregulation of California's electricity market. And then the energy giant played a substantial role in the failure of that very same market. Its market manipulation and fraudulent accounting practices (which led to the belief that Enron was highly profitable when, in fact, it was losing money) led to the failure of California's energy market, with rolling blackouts and the need for state government intervention to stabilize energy supply through the purchase of highly inflated electricity contracts.[8] (The ensuing political fallout also led to the removal of Governor Gray Davis by a recall election.) In widely publicized cases such as these, public policy can lead to the creation, alteration or destruction of markets—all the while the goal being to promote energy security.

There are multiple governmental and non-governmental entities that influence energy supply, and multiple nodes in the energy network that can either facilitate or impede the supply of energy at any particular time. These entities often reflect different value sets, institutional priorities and intent regarding energy development, accessibility, marketing, and pricing. This diversity of perspectives greatly complicates the issue of energy security. While we wish to benefit from the regular supply of energy demanded being available at a reasonable cost to meet commercial and household needs, accomplishing these goals is a large and difficult undertaking that requires a deep understanding of the complex concept of energy security.

Conceptualizing Energy Security

Absolute energy security is not an achievable state, but there is most certainly a continuum along which lies a range of states or conditions between energy security (desired) and energy insecurity (not desired). Each point along the continuum poses its own benefits, costs and risks of failure. Every state or condition poses some form of risk to energy availability and price, as well as their associated political, economic and environmental impacts. Understanding those risks helps us to arrive at a more complete understanding of energy security and the public policies most likely to promote its strengthening over the long term.

Numerous works address the contemporary nature of energy security and insecurity. Sovacool, Kalicki and Goldwyn, and Pascual and Elkind have compiled comprehensive edited volumes examining the myriad aspects of energy security in great detail.[9] For our purposes, a good starting point comes from Daniel Yergin. In his seminal works on energy supply and security, *The Prize: The Epic Quest for Oil, Money, and Power* and *The Quest: Energy, Security, and the Remaking of the Modern World*, Yergin argues that petroleum has greatly shaped, and continues to shape, political and economic realities in most of the modern world.[10] The advent of a new energy age with oil at the center led to a new political paradigm, one driven by the need to maintain ready access to energy resources. In his 1988 article, "Energy Security in the 1990s," Yergin succinctly defines the concept:

> The objective of energy security is to assure adequate, reliable supplies of energy at reasonable prices and in ways that do not jeopardize major national values and objectives ... The focus of energy security concerns is on the shocks—interruptions, disruptions, and manipulations of supply—that can lead to sudden sharp increases in prices and can impose heavy economic and political costs.[11]

Yergin's definition of energy security is broad and inclusive, recognizing the supply and price dimensions, while simultaneously leaving open the door to an array of energy resources, both non-renewable and renewable alike. At the same time, the definition contains elements that are sometimes contradictory, and may therefore work against one another depending on the context of the "major national values and objectives" at play at any point in time.

Yergin's definition of energy security demands a great deal for absolute satisfaction: adequate, reliable, reasonably priced, and consistency with national goals and values. Does the definition ask too much? It is entirely possible that energy supplies will reflect the cost not only of current supplies but also of ensuring future supplies. That objective could be accomplished through regular exploration for new energy resources or through intensive efforts to protect supply chains. If exploration costs rise or costs of maintaining the supply chain increase, then it is likely those costs will be passed on to the consumer in the form of higher energy prices. Is a possibly elevated per unit price for energy something that is "reasonable"? From the position of the free market energy trader, the price increase would be entirely reasonable. Consumers, however, might hold a very different view of the situation and exact a political price from those in political office who enact such a new cost. Given finite resources available in the household budgets of most people, historically sharp increases in energy prices have been viewed typically with deep suspicion and ample frustration.[12]

Yergin's broad definition also points to the role of "values." A unit of energy—a gallon of gasoline or a cubic foot of natural gas, for instance—is fully fungible in any market and holds no single national identity. Once the energy source enters the free market, the energy is simply a commodity to be bought, sold, traded and used by the final purchaser in the chain running from original source through various distribution entities. The energy source might be reasonably priced and represent an excellent source of regular and reliable energy supply. Yet, the energy provider—for example, the leaders of the energy-exporting nation—might also be human rights violators on a regular basis, or the energy might have been pumped from the ground of a politically or socially repressive nation. Would the latter "facts on the ground" matter if the fuel was readily abundant and reasonably priced? It would likely disturb a portion of consumers if they knew that human rights violations or political repression were occurring in an energy-producing nation, and that their energy purchases at least indirectly supported such practices. While one is left with more questions than answers in taking a broad perspective on energy security, the benefit of Yergin's comprehensive definition is that it makes us think a great deal more about what energy security represents in practice, even if it contains potentially contradictory elements.

In "Conceptualising Energy Security and Making Explicit Its Polysemic Nature," Lynne Chester deepens this understanding of energy security in the modern era, illustrating the complex network and often short-lived and rotating nature of political and economic actors in energy markets.[13] She states:

> Twenty-first century access to energy sources depends on a complex system of global markets, vast cross-border infrastructure networks, a small group of primary energy suppliers, and interdependencies with financial markets and technology. This is the context in which energy security has risen high on the policy agenda of governments around the world and the term "energy security" has quietly slipped into the energy lexicon ... An examination of explicit and inferred definitions finds that the concept of energy security is inherently slippery because it is polysemic in nature, capable of holding multiple dimensions and taking on different specificities depending on the country (or continent), timeframe, or energy source to which it is applied.[14]

Chester argues that there are multiple core aspects of energy security.[15] First, it is about the *management of risk*.[16] In order to understand risk, one must understand the probability of various events occurring that have the potential to seriously disrupt energy supply and cause harmful impacts. For example, if a new nuclear power facility were to be located adjacent to a beach, the likelihood of nature's destructive force damaging the power facility leading to environmental or human health risks would need to be considered by power plant designers, by site identification teams, by regulatory agencies focused on environment and health, and by emergency first responders. Risk calculation and management require the identification of all things that reasonably might disrupt energy supply or delivery to the end-user, and contribute to adverse environmental or human health outcomes.

Second, Chester observes that one must also consider a country's (or continent's) energy use mix, which includes many dimensions, including locally available resources and the world's increasing focus on sustainability.[17] The US energy bonanza focusing on natural gas and shale oil is often framed by elected officials and industry leaders as an excellent example of the maximization of energy security through increased reliance on domestic energy sources.[18] Advocates for renewable energy make a similar argument about the desirability of domestic energy sources that are also less environmentally harmful.

Third, Chester's model is action-based, incorporating the idea of *"strategic intent"* in the pursuit of energy security.[19] Through promotion of energy security, a nation can enhance its position and capacity to act in relation to markets and competitor nations by anticipating looming threats. This idea does not necessarily imply conflict. It can also feature cooperative mechanisms for enhancing a nation's energy security position.

Fourth, "...energy security has a temporal dimension. The risks or threats to supply differ across short, medium, and long term horizons."[20] The challenge that emerges for energy industry leaders and energy policymakers is to be able to adjust quickly to changing conditions and to be able to discern the time and resource commitment needed to manage particular threats to energy security, realizing all the while that there is no desired end-state but rather only continual ongoing risk management being possible.[21]

In contrast to this framework, Christian Winzer, in "Conceptualizing Energy Security," offers a narrower, yet also a deep approach to defining energy security.[22] Unlike Yergin and Chester, who open the door to myriad policy values that contribute to wide-ranging policy frames for energy security, Winzer suggests "narrowing down the concept of energy security to the concept of energy supply continuity. This reduces the overlap between the policy goals of energy security, sustainability, and economic efficiency."[23]

Winzer's model of energy security still reflects the complex and multidimensional nature of the concept, even when narrowly defined. He identifies three major sources of risk: technical, human and natural.[24] Technical risks that might have an impact on energy security are related to the mechanical soundness of energy-related production equipment. Human risk sources are highly varied, ranging from terrorism and purposeful sabotage to the use of energy as a tool in political and diplomatic relations. Natural risks refer to the variation and inevitable decline of energy resources due to natural disaster or depletion.[25]

A second dimension to Winzer's model is termed "scope of the impact measure." There are four key elements to this: "continuity of commodity supply," "continuity of service," "continuity of the economy," and "environment and society."[26] Continuity of commodity supply is a key concern for those who conclude that petroleum, natural gas, and coal supplies are finite and will face eventual depletion. Continuity of service refers to "the continuity of the price and the availability of energy services such as heating, lighting, communication or transport."[27] The continuity of the economy refers to impacts that can threaten the standard of living in a region or nation. An example of this is the oil spill in the Gulf of Mexico in 2010 that led to the release of over 200 million gallons of oil into the waters of the Gulf. The size of the spill was so significant that it continues to have an adverse impact upon fisheries and Gulf state tourism that could weaken the economy of the region for a long time.[28] The fourth element relates to energy security risks that threaten "human safety and environmental sustainability."[29] The nuclear disaster at Chernobyl in 1986 is a poignant example of the impact of energy security risks that threaten human health and produce long-lasting environmental disaster.

The third dimension of Winzer's model of energy security involves the severity of supply disruption impacts, such that "the severity of threat increases with the speed, size, sustention, and spread [of impact] as well as with decreasing singularity and sureness of the impacts."[30]

Winzer's model of energy security challenges broader understandings of the concept, such as those offered by Yergin and Chester, while still offering an understanding of energy security that is complex in its various dimensions. At the same time, Chester's and Winzer's models of energy security are more empirically driven models of energy security. Both models provide opportunities to consider multiple dimensions of energy security and to describe and explain noteworthy energy security outcomes. Winzer's model provides a comprehensive framework with which to assess the origin and nature of energy security impacts. Chester's model offers a useful taxonomy of energy security dimensions and impacts, emphasizing both the qualitative and normative aspects of energy security to highlight the complex political, economic, and social dynamics undergirding energy security issues. These two approaches, however, do not fully take into account the broader environmental elements that can (and should) be included in the idea of energy security.

Over the past few decades, as a new energy paradigm has emerged with environmental concerns at its core, there is an increasing emphasis in energy security policy in the type of fuels used. When considering the utility of a particular energy source, it is increasingly necessary to consider whether the use of the source is environmentally and/or socially responsible:

- Will there be a carbon footprint that contributes to climate change?
- What are the costs and negative externalities associated with a particular energy source?
- How will the use of an energy source impact the health of communities, the young, the elderly?

Benjamin Sovacool has emerged as one of the leading proponents of this analytical approach. His earlier work argues that the systematic comparison of energy sources in the United States consider the following criteria:

(a) technical feasibility, meaning that such systems must be commercially developed and available to enter the American energy market; (b) cost, in terms of whether their use would increase or decrease electricity prices for consumers; (c) negative externalities, in terms of their impact on human health and the environment; (d) reliability, in terms of how dependable such technologies are at generation and delivering electricity; and (e) security, or how safe and immune such technologies are from attack or accident.[31]

Based upon these criteria, Sovacool concluded that, "The wisest energy strategy for the United States—in terms of cost, environmental and potential … is to invest in long-term energy demand reduction through the increased deployment or improved performance of energy-efficient equipment. On the supply-side, using smaller, decentralized units such as wind turbines, combined heat and power systems, biomass generators, and solar heating and photovoltaic systems is a much better strategy."[32]

Sovacool has more recently argued for an even greater expansion of the concept of energy security. In the article "Evaluating Energy Security Performance from 1990–2010 for 18 Countries," Sovacool and his co-authors apply a broadly inclusive definition to see how countries "equitably provide available, affordable, reliable, efficient, environmentally benign, proactively governed and socially acceptable energy services to end-users."[33] Sovacool has pointed out in his *Routledge Handbook of Energy Security* that the definitions of energy security are numerous, 45 by his count, but he rejects narrow understandings of energy security, arguing that:

> It is no longer appropriate to envision and practice energy security as merely direct national control over energy supply, and instead necessitates carefully cultivating respect for human rights and the preservation of natural ecosystems along with keeping prices low and fuel supplies abundant.[34]

In conceptualizing energy security in this way, beyond the issue of supply continuity, Sovacool has been able to incorporate the many value-based dimensions that focus attention on planning for the future to avoid harmful impacts.

A Framework for Analysis

Building upon the work of Sovacool and the conceptualizations described above, we address the problem of energy security in the United States and the policies employed to achieve this goal. We offer a broad understanding of energy security that is based, first of all, on the idea that the United States' national energy system involves four core elements:

- *Public institutions*: elected and appointed officials in domestic and international governing bodies, bureaucracies, non-governmental organizations operating inside the power and energy sector, universities engaged in energy research.
- *The market/business sector*: industrial and commercial energy consumers, power companies, system operators, energy resource firms, technology developers and manufacturers.
- *Citizen-stakeholders and civil society*: individuals, interest groups, secular nonprofit organizations, religious organizations, and professional organizations.
- *Technology*: the fuel sources and technologies that make energy available.[35]

Each of these areas is a "driver" of energy security and energy policy, while at the same time each is affected by developments involving the other areas. In addition, actors in these sectors operate on the basis of uneven levels of knowledge, are concerned with different time frames, and in many cases are responsive to diverse actors in the political, public or private sector. This variety of contexts within which each potential driver of energy security and energy policy operates can lead to quite different understandings, or mutual misunderstandings, regarding choices and risk, or too often to complacency with regard to difficulties in determining risk at any one point in time.

Second, our concept of energy security recognizes, as do Yergin, Chester and Winzer, that energy security involves primarily concerns about supply, access and affordability, consistent with larger national goals. But at the same time, we attach great importance, as Sovacool does, to the idea that energy security reflects a multitude of concerns and objectives, including evolving community, national, and global environmental values.

Given the foregoing discussion, we posit here that energy security includes the following components, and that there are distinct tiers of importance with respect to the goals of energy security:

Abundance of energy supplies. American society and the US economy demand that energy supplies are sufficient to be called upon at any time, from different sources, to provide electric power, heat and liquid fuels for transportation. *Abundance* means that there is enough energy for everyone's needs, so that everyone can access its benefits. It can come in the form of finding new or producing more from known energy sources, and in the form of greater energy efficiency (doubling the fuel efficiency of cars effectively doubles the supply of gasoline).

Reliability of energy sources. *Reliability* means that governments, industries and individuals can remain assured that energy supplies cannot and will not be easily disrupted—that the energy will be there when people need it to carry on their work and live their lives in relative comfort. Robust reliability would mean that political conflict and volatility in the Middle East would not result in price spikes due to temporary cuts in oil supplies. Nor would a natural disaster such as

Hurricane Katrina be able to seriously affect production capacity. Reliable power means fewer periodic or rolling blackouts due to high electricity demand, electrical grid inefficiencies, or technical failures that cascade out of control as did the power outage that struck much of the Northeastern United States and Southeastern Canada in 2003.

These first two aspects of energy security occupy the top tier because if the lights go out, the production lines stop, or gasoline lines develop, this event is considered unacceptable among energy users, energy suppliers and government agencies. Those who suffer a negative impact by such shortages tend to demand immediate governmental action and relief if serious loss is suffered, while energy providers, with the urging and sometimes the assistance of government, promise to restore services and supplies as quickly as possible.

Affordability of energy sources. This component of energy security means that long-term prices are kept manageable and are not subject to high inflation. *Affordability* allows businesses, governments and individuals to predict future costs accurately and to access funds that would otherwise be spent on energy, using them for other investments. It also means that short-term price volatility would be diminished, allowing energy users to avoid the potentially large costs that come from a rapid spike in energy costs. This aspect of energy security includes not only prices, but also the time horizon in which end-users can and do find a way to accommodate themselves to higher prices (e.g., purchase of more fuel-efficient vehicles). When markets are volatile and sharp price increases occur, affordability becomes a more immediate element of energy security. Affordability sits just below abundance and reliability in order of relative importance. People expect prices to rise over time; moreover, the term "affordable" includes a wide range of interpretations depending upon one's circumstances.

Diminishing the human and environmental impacts of energy production and use. The growing awareness and understanding of the downstream impacts of our energy choices has led to a growing effort to diminish these consequences: climate change, air pollution from NOx, SOx and particulate matter, oil spills, the despoliation of land and water that result from the methods by which the United States acquires oil and coal, the deaths and public health effects of acquiring and burning fossil fuels—all of these negative externalities have been shown to result from the widespread use of fossil fuels. This aspect of energy security, which can be described as including the goals of environmental *cleanliness and sustainability*, is afforded a longer time horizon than the previous three. Consumers, firms, communities and government agencies tolerate a certain level of human and environmental damage in the operation of our energy system. Mine disasters, oil spills, coal impoundment, dam breaches, greenhouse gas emissions, particulate matter emissions, higher than expected rates of lung cancer occurring in communities where coal is mined or burned in power plants—these adverse impacts are generally considered to be an expected part of producing and using energy.

Diversification of energy sources. This goal includes diversification of both fuel types and geographic sources of energy. *Diversity* of supplies and suppliers supports all of the energy security elements described above, while also serving the objective of mitigating national and global security vulnerabilities. Success in achieving diversification, ideally, would greatly reduce dependence upon oil for transportation and dirty fuels for electricity generation. With respect to diversifying suppliers of oil, away from OPEC countries in particular, this is commonly referred to as "energy independence," a term which should not be equated with total self-sufficiency. More properly understood it means diminishing or ending the role of oil as the world's ultimate strategic commodity, making it just another commodity in which to trade, and thereby mitigating the national security and economic impacts of political instability, terrorism and military conflict in the Middle East.[36] This objective maintains an even lower priority than the other specified goals, as it is based upon long-term objectives that are desirable, but inherently elusive. However, it is probably the most important element of any effort to achieve a long-term, resilient state of greater energy security. Continual changes in supply and demand, prices, technologies, and policies, along with conflict and violence around the globe, all routinely present challenges to governments, firms and consumers (energy forecasting is notoriously difficult). Diversification of energy sources can significantly diminish the importance and impacts of disruptions and surprises relating to any one fuel source or supplier.

On top of all this, each of these goals has to be pursued, as Daniel Yergin put it, "in ways that do not jeopardize major national values and objectives."[37]

This is where the critical role of politics and policy come into active play. To help achieve these objectives, policies are adopted to meet one or more of these goals. On the domestic front, drilling for oil and gas offshore and on public lands; regulations to require scrubbers at power plants to prevent release of pollutants into the air; tax deductions and credits for oil, gas, solar, wind, biomass and other fuels and technologies; and federal indemnification for nuclear power disasters are all ways in which policy is designed to achieve various aspects of energy security. With regard to foreign policy, the alignment of energy and national security interests includes the effort to maintain a robust supply chain for oil, enhance stability and diminish conflict among oil-producing countries and regions, develop a more globalized natural gas market, and counter the ability of states such as Russia and Iran to use energy as a tool to achieve other political ends, all while acknowledging the need for a long-term strategy to address the environmental and climate impacts of continued fossil fuel use.[38]

There is, however, no single policy approach and there is likewise no single energy source that is sufficient to meet all the goals we have as a society. In order to make gains regarding one objective, we have to accept some setbacks with regard to others. This is the *American energy security dilemma*. Efforts to increase security in one realm often result in a decrease in security in another realm. US energy security involves multiple goals that are at times contradictory, and so the policies adopted in their service embody tradeoffs that cannot be avoided. In other words, the pursuit of energy security, broadly understood to include multiple technical, social, economic and political objectives, is inherently self-contradictory in nature.

For example, in order to increase the abundance of oil, deepwater drilling in the Gulf of Mexico is pursued, but oil spills such as the Deepwater Horizon accident have diminished the environmental health of the region and caused many individuals the loss of their livelihoods. While solar and wind are seen as good alternative energy sources with respect to human health and environmental impacts (though not direct substitutes for oil), their availability—both the volume of production and the fact that solar and wind power are intermittent resources— means that they cannot provide (at least not at this time) a sufficient supply of electricity to meet the nation's demands. Moreover, solar and wind are generally not quite cost-competitive with coal and natural gas power plants at this stage of their development in most markets. (However, this price differential is rapidly shrinking, and "grid-parity" seems to be an attainable goal in the not-too-distant future.)

It is important to note that the energy security dilemma is affected further by the fact that such security is temporal—circumstances change over time, while values evolve and risk perceptions can change along with those new circum- stances and altered value orientations. Through both the political process and through marketplace choices made by consumers, the United States continually revisits its aims and methods of establishing a secure energy supply. Whatever energy choices are made—either collectively or individually—there are corre- sponding tradeoffs and levels of risk that are shouldered by the society at large. Our risk profile might look very different if we drew on domestically-produced of energy sources, such as wind, biomass, coal, and geothermal rather than pet- roleum resources being transported via tanker from a distant port located in a region often characterized by political and social turmoil.

An added element of energy security, beyond that of evolution and change, is inevitable ambiguity. Energy security requires access to high quality current information and data sources on energy demand, energy type, and associated risk. The ability to access these dimensions of energy security has improved greatly in the Information Age. Data on American energy supplies and energy security failures is now abundant and easily located, and much of it can be down- loaded from government web portals. Still, corporate data on energy risk is proprie- tary and tends not to be as easily accessible, while the accuracy and completeness of international data sources is highly dependent on the information provider. Data completeness is also hampered by the fact that many dimensions of energy security are not easily measured and monitored. The issue of trust and cooperative energy resource development is one of the key areas where energy security assessment suffers from incompleteness. A sudden shift in policy, such as Russia's shutting off of natural gas supplies to Ukraine or its annexation of Crimea, are significant events not easily predicted in terms of timing and implication for energy supplies, markets, and associated tradeoffs.

In this context of an energy security dilemma that is further impacted by change and uncertainty, what American energy policy does is to target a variety of energy sources so that, to some extent, all the goals of American energy security are being pursued simultaneously, even if different policies may work at cross-purposes with one another or get uneven levels of priority. On the surface this patchwork quilt of policies can seem to be a self-contradictory, unwise approach to energy. However, the American energy system is one that constantly evolves, based upon changing values and interests, rival interest groups, competitive commercial production and marketing companies, both old and new technologies, diverse fuel sources, partisan politics and active policy advocacy, and newly arising market opportunities and barriers to once open access. Perhaps an apt analogy to suggest order out of this apparent chaos is that of a kaleidoscope. There are numerous individual parts that affect how energy is produced, acquired, used and mitigated in its impacts. And while the different parts can collide with each other, in the end there is a degree of order and stability in the system as the parts also complement each other—just as a kaleidoscope displays a continually changing coherent whole out of myriad parts. Policy itself is a key part of this system, helping to structure the market, promote technology development, diminish national security and environmental impacts, and achieve all the aims of energy security that may be attainable at any point in time. To this end, specific policy tools can be at once contradictory and complementary to one another.

The following chapters examine this concept of energy security and its myriad impacts by reviewing the evolution of American energy policy, examining the interplay of markets and public policy, describing the ongoing transformation of the US energy landscape, and characterizing the global context in which energy security is being pursued. Chapter 2 looks at energy security in terms of the change in American values that have expanded our understanding of energy security over time. These values have been made evident in both the adoption of public policy and in the actual consumption of different types and amounts energy resources. Chapter 3 looks at the American approach to energy resources as both a private good and one that provides a shared, public benefit. It considers the interaction between public policy and markets in establishing what we call the "marginal social costs" of the American energy profile. Chapters 4 and 5 address the contemporary landscape of US energy policy in the pursuit of energy security. Chapter 4 examines the ongoing transformation (both intended and unplanned) with regard to fossil fuels, focusing on oil and gas production as a result of the fracking revolution, and looks at the growing efforts to diminish either the use of coal or mitigate its known harmful impacts. Chapter 5 looks at this same transformation in the realm of renewable energy, exploring the growing use and impacts of biofuels, along with the growth of solar and wind power across the United States. Chapter 6 considers these domestic developments in the larger context of the global dimension of US energy security, examining global institutions and a number of regional energy issues that impact the pursuit of the country's energy objectives. Chapter 7 offers our concluding thoughts.

Notes

1 Fareed Zakaria, *The Post-American World*, New York: W.W. Norton and Company, 2008, p. 1.
2 Rajeev Dhawan and Karsten Jeske, "How Resilient is the Modern Economy to Energy Price Shocks?" *Economic Review*, 3, 2006, pp. 21–32; and Jame DiBiasio, "Where the Danger Lies," *Asian Investor*, December 1, 2010.
3 See David L. Sills, "The Environmental Movement and its Critics," *Human Ecology*, 3(1), 1975.
4 Friedrich Hayek, *The Road to Serfdom*, Chicago, IL: University of Chicago Press, 1944.
5 Milton Friedman, *Capitalism and Freedom*, Chicago, IL: University of Chicago Press, 1962.
6 Peter Grossman, *US Energy Policy and the Pursuit of Failure*, Cambridge: Cambridge University Press, 2013, p. 30.
7 Toby Bolsen and Fay Lomax Cook, "Public Opinion on Energy Policy, 1974–2006," *Public Opinion Quarterly*, 72(2), 2008.
8 Jacqueline L. Weaver, "Can Energy Markets be Trusted? The Effect of the Rise and Fall of Enron on Energy Markets," *Houston Business and Tax Journal*, 4, 2004.
9 Benjamin Sovacool, ed., *The Routledge Handbook of Energy Security*, Abingdon: Routledge, 2011; Jan Kalicki and David Goldwyn, eds, *Energy and Security: Strategies for a World in Transition*, Washington, DC: Woodrow Wilson Press, 2013; Carlos Pascual and Jonathan Elkind, eds, *Energy Security: Economics, Politics, Strategies and Implications*, Washington, DC: Brookings Institution Press, 2009.
10 Daniel Yergin, *The Prize: The Epic Quest for Oil, Money, and Power*, New York: Free Press, 2009; and Daniel Yergin, *The Quest: Energy, Security and the Remaking of the Modern World*, New York: Penguin Press, 2011.
11 Daniel Yergin, "Energy Security in the 1990s," *Foreign Affairs*, 67(1), 1988, p. 111.
12 Scott McCulloch, "One Million Can't Afford to Heat Their Homes: Record Fuel Poverty Levels after Energy Price Hikes," *Daily Record*, September 16, 2011.
13 Lynne Chester, "Conceptualising Energy Security and Making Explicit its Polysemic Nature," *Energy Policy*, 38, 2010, pp. 887–895.
14 Chester, p. 887.
15 Chester, p. 889.
16 Chester, p. 889.
17 Chester, p. 890.
18 Garrick B. Pursley and Hannah J. Wiseman, "Local Energy," *Emory Law Journal*, 60(4), 2011, pp. 877–969; Abby Schachter, "Energy Independence and its Enemies: The Bounty of Shale-oil and the Environmentalist Forces that Want to Keep it Buried," *Commentary*, 133(6), 2012, pp. 24–29.
19 Chester, p. 890.
20 Chester, p. 890.
21 Chester, p. 893.
22 Christian Winzer, "Conceptualizing Energy Security," *Energy Policy*, 46, 2012, pp. 36–48.
23 Winzer, p. 36.

24 Winzer, p. 37.
25 Winzer, p. 27.
26 Winzer, p. 38.
27 Winzer, p. 38.
28 Peter Grier, "Gulf Oil Spill: The 51 Minutes that Led to a Disaster," *Christian Science Monitor*, May 26, 2010.
29 Winzer, p. 38.
30 Winzer, p. 37.
31 Benjamin K. Sovacool, "Coal and Nuclear Technologies: Creating a False Dichotomy for American Energy Policy," *Policy Sciences*, 40(2), 2007, p. 102.
32 Sovacool, "Coal and Nuclear Technologies ...," p. 118.
33 Benjamin K. Sovacool, Ishani Mukerjee, Ira Martina Drupady, and Anthony L. D'Agostino, "Evaluating Energy Security Performance from 1990–2010 for Eighteen Countries," *Energy*, 36, 2011, p. 5846.
34 Benjamin Sovacool, "Introduction: Defining, Measuring and Exploring Energy Security," in Sovacool, ed., *The Routledge Handbook of Energy Security*, Abingdon: Routledge, 2011, pp. 3–6, 9.
35 Christopher A. Simon, Christine Taylor, and Theodore Batchman, "Renewable Energy Policy Innovation and Interdisciplinary Education: Cross-Discipline Instruction in Engineering, Economics, and Political Science," presented at the Annual Meeting of the Midwest Political Science Association, Chicago, IL, 2008. See also Aaron Wildavsky and Karl Dake, "Theories of Risk Perception: Who Fears What and Why?" *Daedalus*, 119(4), 1990, pp. 41–60.
36 Gal Luft, "Saudi Pushers, Energy Rehab," Middle East Strategy at Harvard (MESH), John M. Olin Institute for Strategic Studies, September 1, 2009, http://blogs.law.harva rd.edu/mesh/2009/09/saudi-pushers-energy-rehab/ (accessed September 26, 2013).
37 Daniel Yergin, "Energy Security in the 1990s," p. 111.
38 Jan Kalicki and David Goldwyn, *Energy and Security: Strategies for a World in Transition*, Washington, DC: Woodrow Wilson Press, 2013.

2

VALUES, CHOICES AND NEEDS

The public affairs realm of US energy policy, if it were to be presented as a visual map, might look quite haphazard and appear to be continually in flux. After all, there are a great number of policies by which the United States seeks to achieve energy security, and a large number of actors are involved in the development and implementation of the nation's energy policy. This is a circumstance that is at once *unavoidable* (we live in a representative democracy with many well-organized interests seeking representation through a myriad of governmental entities), *indispensable* (the energy landscape is always changing and we need to respond to it with flexibility and dexterity), and *problematic* (it therefore becomes difficult to arrive at a focused energy security strategy beyond an "all of the above" approach).

At the same time, the analogy of the kaleidoscope suggests a degree of order, or at the very least, the fact that a series of agreements, or settlements, are continually being produced in the nation's policy process. This process, by which the United States seeks to achieve maximal energy security, is a reflection of these core elements:

- *Societal values*, which continually evolve.
- *Energy needs*, which continually grow.
- *Feasible policy choices* that lie within existing political and economic contextual boundaries, a policy space that offers opportunities and imposes constraints that are themselves shaped by the values, interests and needs of the principal actors working in the energy policy realm.

These core elements—values, needs and feasible choices—all interact with one another to produce the system by which the United States produces, acquires and consumes its energy.

Values and Choices

Developing an understanding of what energy security means, and then pursuing it, is a function of many, sometimes conflicting, forces, sometimes acting in unison and often moving in opposing directions. These forces are informed and driven in part by values, which play a large role in shaping private and public choices. The multiple goals of US energy security are themselves values—a reflection of what this country, as a collective entity, cares enough about in the energy sector to make choices involving a great deal of time, effort and money in their pursuit. Moreover, this process undergoes ongoing revision, as the institutions and policies that are subsequently developed to pursue our nation's choices are not only shaped by these values, but can also magnify the centrality of particular values and relegate other values to the sidelines. A close look at how our nation's values concerning energy have evolved is an important element of understanding of energy security in our contemporary world.

The Evolution of Energy Values and Energy Policy in the United States

Aiming for, and Achieving, Abundance and Affordability

In the United States, as it entered the 20th century, when markets for oil and electric power were just starting to develop, the prevailing energy security values that came to be reflected in policy were those of pursuing *energy abundance*, while also extending access to energy at *an affordable price* to all of the nation's citizens, even the least benefitted members of society—in particular, those living in relative poverty, in distant rural corners of the land, enduring the adverse consequences of ongoing energy poverty. At the same time, these social equity and economic opportunity goals were to be pursued without abandoning the nation's longstanding commitment to free market capitalism.

Energy policies adopted in the early decades of the 20th century embodied these multiple, sometimes competing goals, particularly during the Great Depression when providing energy to the underserved became a principal governmental priority. The political climate and circumstances of the time offered the window of opportunity by which broad national energy goals could be developed and pursued by the federal government. FDR's presidency both responded to and cemented the idea that the federal government would take greater responsibility for the individual welfare of citizens, and would provide direct relief to people in the midst of economic downturn. To that end, national energy policies such as the *Rural Electrification Act of 1936*, the *Bonneville Project Act of 1937*, and the *Tennessee Valley Authority Act of 1933* sought to extend the reach of the electric power grid to communities that had been largely left out of the electric energy market. Other actions, such as the breakup of the Standard Oil Corporation in 1911, the enhanced regulation of the oil industry in the 1930s and 1940s to stabilize supplies and prices, and the passage of the *Public Utility Holding Company Act of 1935* collectively sought to protect the market by diminishing the reach and power of key companies in the oil and electric power sectors so that the supplies and prices of energy resources could become more stable and entail at least some degree of competition.

These policies were both a reflection and an extension of the paradigm built around a supply-focused view of energy security. Adequate supply was to be measured by more than total quantity of energy, but also by the geographical distribution and the affordability of said quantities. These policies began the process of bridging the gap between the aim of fair access and the sole reliance on free market mechanisms to allocate energy resources. Public policy helped to expand the conception of energy security. Energy became more than a strictly "marketable good" that existed solely in the private realm and that could be rivalry-oriented and excludable, and toward the status of being a "marketable public good" with the attendant elements of public interest calculations being present in policy debates. In this sense, we are not using the term "public good" in the strictly orthodox economic sense—a good that is both non-rivalrous and non-excludable, such as free radio or clean air. Rather, we are looking toward the view that suggests "public goods are socially defined and constructed according to what is perceived as a 'public need', rather than containing certain inherent characteristics of non-excludability and non-rivalry."[1] Based upon this understanding, we argue that the public and social benefits of energy came to be increasingly incorporated into the logic and formation of public policy, and that energy security itself has come to be understood as a public good.[2] The public policy results emerging from this period were remarkable in many respects. In the United States the values of abundance, affordability and access were reflected in both public policy and in producer and consumer choices, such that by mid-century the infrastructure for supplying electric power and transportation fuel reached just about every remote corner of the country.

When World War II ended, the United States emerged not only as a political, economic and military superpower, but an energy superpower as well. In the postwar era, up until the 1970s, the US benefitted from a circumstance in which it was almost entirely energy self-sufficient, while being one of the world's largest producers and exporters of oil. Both domestically and abroad, energy resources were available in considerable abundance, a condition that kept energy prices low and fueled rapid growth in energy consumption. Oil, priced at no more than \$2–3 a barrel, came to increasingly displace coal in electricity production, as its ease of acquisition and use offered considerable advantages over coal.[3] At the same time, coal and natural gas remained both plentiful and relatively inexpensive. Coal remained important to electricity production, while coal exports also helped keep the industry productive and profitable. Natural gas prices were regulated and maintained at a low level, while domestic markets for natural gas were largely centered on cooking and heating uses, and less oriented toward use in electricity generation. And of course, gasoline was quite cheap, staying well below 50 cents a gallon for many years.[4] What consumers saw was that energy in all forms was just plain cheap. Plentiful, affordable energy fueled remarkable economic growth in the US (and other industrialized countries). It appeared as if the United States had achieved its energy security goals of abundance and affordability for all with little effort beyond the discovery of ever more vast supplies within the country and abroad. This is what the country valued with respect to energy, and these aims were part of the larger objective of maintaining American prosperity and the nation's global military and political power.

Energy policy during this period reflected the seemingly unquestioned belief that supplies were virtually endless and energy prices would always stay low as a consequence, fueling both domestic and global economic growth as far out into the future as one could imagine. A system of federal and state controls on oil suppliers kept domestic producers and consumers alike pretty much satisfied. For example, the Texas Railroad Commission was a particularly powerful entity, exercising effective control over domestic supplies and prices. They operated with the agreement of producers who benefitted from high prices—compared to international markets—and consumers who received reliable supplies at affordable prices. The US Congress and President Eisenhower further acted in concert to protect the domestic energy market. Congress passed the *Reciprocal Trade Act Amendments of 1955*, legislation that empowered the President to restrict the import of a commodity if the imports harmed national security. In 1959 the President invoked the import restriction clause and limited oil imports. With respect to the increasingly important oil-producing countries in the Middle East, a policy that ensured the continuing flow of oil was, for the most part, sufficient, ensuring an energy surplus in Europe and the developing world to stimulate economic growth.

The other major development in this period was the birth of the nuclear power industry. Congress supported the emerging industry, which could potentially offer even greater supplies to an already flooded electricity energy market. The *Atomic Energy Act of 1954* began the development of a civilian nuclear power industry, and the *Price-Anderson Act of 1957* provided federal indemnification to nuclear plant owners in the event of major disasters resulting from the operation of civilian nuclear reactors providing electrical power.

Looking for Reliability

Circumstances soon changed, of course, in this highly favorable global and national energy scenario. The vulnerabilities in the American energy paradigm first became apparent in 1965 when a massive blackout in the Northeast began in New York and affected six surrounding states and the Province of Ontario in Canada. That event left 30 million people entirely without power. All of a sudden, the liabilities associated with ever greater size—producing and consuming more electricity in bigger power plants feeding sprawling suburbs and growing factories—revealed themselves in dramatic fashion to producers and consumers alike. The prevailing assumption, largely unquestioned or even unacknowledged, that the electrical grid would always work was cast into serious doubt.

The exposure of vulnerabilities in the electric power sector was soon largely repeated in the oil sector. In the early 1970s the falling level of US oil output alongside rising demand, the growing pressure on global oil supplies and prices, and then most importantly, the oil embargo of 1973–4 (see Box 2.1) demonstrated how the widespread growing reliance on oil (particularly on oil from the Middle East) had created huge liabilities and costs for consumers all over the world. In the United States, where energy consumption had grown exponentially since 1945, the disruptions were massive. Accurately termed an "oil shock," the embargo caused oil prices to quadruple, sharp inflation, and the proliferation of gas lines in the nation's major population centers.

BOX 2.1 THE OIL EMBARGO OF 1973–74

The Organization of Petroleum Exporting Countries (OPEC) was formed in 1960 by Iran, Iraq, Saudi Arabia, Kuwait and Venezuela (later joined by several other countries, and currently consisting of 12 members), in response to the dominance of the major Western oil companies of the day, the "Seven Sisters" that largely controlled global oil supplies and prices. OPEC describes its mission as one designed to "coordinate and unify petroleum policies among Member Countries, in order to secure fair and stable prices for petroleum producers; an efficient, economic and regular supply of petroleum to consuming nations; and a fair return on capital to those investing in the industry."

In response to the outbreak of the Yom Kippur War in 1973, and the subsequent supply of US arms to Israel, a subgroup of OPEC, the Organization of Arab Petroleum Exporting Countries (OAPEC), proclaimed an oil embargo against all countries supporting the state of Israel. Subsequently, OPEC raised the price of oil from its members and cut production. The United States faced significant impacts from these actions, as it was the leading supporter of Israel. The result over the next few months was a quadrupling of oil prices from $3 to $12, the beginning of a vast accumulation of wealth among oil producing countries, a realization that the political and economic power of the United States and the West had been weakened, and the assertion of OPEC and its member states in taking greater control over global oil supplies and prices.

The oil embargo and rapid price increase led to an equally rapid change in the political climate in the United States. The turmoil that ensued, with people lining up their cars to purchase gasoline at higher prices than they had ever paid, produced a sense of vulnerability among policymakers and the public. No longer were the United States and its allies as secure as they had believed, and something had to be done about it. An editorial from *The New York Times* in November of 1973 captured this sentiment, stating that, "The oil embargo they have launched is an act of both political and economic warfare ... the Arabs must be disabused that they can go on waging economic war with impunity." The *Times* further reflected the growing awareness that, "Strong conservation measures, combined with all-out efforts to develop energy resources and new technologies, are desirable in themselves ... the current crisis is compelling action to develop new energy sources that in any case would have to be taken by the end of this century."[5]

These developments prompted a change—an expansion, really—in American energy security goals. Americans came to value not only abundance and affordability, but two additional aims came to the fore—namely, *reliability* and *diversification*. Reliability involved the assurance that energy supplies would be available when called upon, something that had come to be taken for granted in the previous period of vast energy surpluses. Diversification involved the dual elements of reducing America's reliance on foreign energy sources (achieving geographical diversity of supplies away from such a heavy reliance on the Middle East), and diminishing the role of oil for civil and commercial transportation and for electricity generation (finding different fuel sources to meet the country's needs). Moreover, the achievement of diversification would itself promote the aims of abundance, affordability and reliability by expanding the scope of energy sources and types of fuels to be produced and consumed. The result of this systematic rethinking of energy policy was the adoption of a series of new energy-related public policies to address these serious deficiencies, as well as the creation of a federal Department of Energy to develop and coordinate a comprehensive national energy strategy and policy.

Major policy measures were adopted at this time to limit the impact of the country's unhealthy dependence on oil. In response to the embargo and the growing realization about American vulnerabilities, Congress passed legislation such as the *Energy Policy and Conservation Act of 1975* (EPCA), and the *National Energy Act of 1978*. These statutes included several measures intended to both increase energy supplies and to promote energy efficiency. The EPCA established a mandate by which cars and light trucks sold in the United States were required to meet federally-designated fuel efficiency targets, known as Corporate Average Fuel Economy, or CAFE, standards. These provisions went into effect in 1978, ramping up from 18 mpg to 27.5 mpg by 1985. The standard remained at this level for 20 years, but the target for fuel efficiency was then significantly increased via both legislation and regulation, with a mandate to achieve an average of up to 54.5 mpg by 2025.[6]

Another major element of EPCA was the creation of the Strategic Petroleum Reserve (SPR) in 1975. Designed to mitigate temporary supply disruptions, the

SPR would allow for the purchase of oil and its placement in the reserve when prices were low and stable, and permit its release when markets experienced turmoil and rapid price increases. By making itself an energy market actor, the US government could possibly diminish the chance that disruptions or threatened disruptions of supplies would trigger price spikes in the oil market. (The most notable recent sale of oil from the SPR was in the aftermath of Hurricane Katrina.) Maintaining a strategic reserve of oil is also a treaty obligation that the US maintains as part of its membership in the International Energy Agency, which was established in the wake of the oil embargo. As of September 2015, the SPR contains 695 million barrels of oil—with a total capacity of 713 million barrels—enough to replace all crude oil imports for about 137 days, based on the total net imports in 2014.[7]

To help offset the use of oil, the country also looked to the promise of ethanol. Congress approved subsidies to ethanol producers in 1978 with the *Energy Tax Act* (a part of the *National Energy Act*), providing a tax exemption of $.40/gallon. From this time through 2011, when the subsidy was eliminated, producers of ethanol received anywhere from 40 to 60 cents a gallon, as subsequent legislation— the *Alternative Motor Fuels Act*, the *Energy Policy Act of 1992*, the *Energy Policy Act of 2005*, among others—both extended the subsidy and revised the amount of the tax credit.[8] (The subsidy for the production of advanced biofuels such as cellulosic ethanol was not phased out in 2011, and remained in force in the amount of up to $1.01 per gallon through 2014. Advocates of the advanced biofuels tax credit have lobbied extensively to extend the measure.[9]) In addition to providing tax incentives, Congress also protected the domestic market by placing a $.54/gallon tariff on ethanol imported from other countries. This measure, in effect from 1980 through 2011, was largely designed to curb imports of sugarcane-derived ethanol produced in Brazil.

With respect to the electric power sector, the twin goals of reliability and diversification resulted in noteworthy changes as well. Though overshadowed by gas lines and the rising price of oil, the vulnerabilities of the electric power sector were also significant. In the aftermath of the 1965 blackout, utilities took direct action to prevent similar occurrences. The National Electric Reliability Council was established in 1968, along with several regional reliability councils, to make the nation's grid more robust and reliable, developing standards and fostering greater coordination among the nation's utilities. At the same time, new metering and monitoring technology was deployed over time to improve grid reliability substantially. National policy also sought to make changes in the industry, and major legislation was included in the *National Energy Act*. The *Power Plant and Industrial Fuel Use Act* mandated that utilities move away from using oil as a source of their fuel. This dictate coincided with the interests of utilities, which could acquire coal domestically at a lower cost than oil. At the same time, the *Natural Gas Policy Act* sought to diminish the regulatory controls on both the price and transportation of natural gas, an action that would allow for greater production. The *Public Utilities Regulatory Policy Act* (PURPA) allowed qualifying electrical generation facilities not owned by utilities to sell power to the utilities, and it forced the utilities to purchase the power in question. This policy allowed for expansion of the electricity supply by new market actors (merchant generators and independent power producers), and sought to encourage the timely growth of wind power and other renewable energy supplies (though over time the biggest expansion came in the form of natural gas plants). Eventually, PURPA helped to set in motion a major restructuring of the nation's electric power industry.

Enter Environmentalism and Concern for Sustainability

At about the same time that the goals of reliability and diversity came into focus, in the 1960s and 1970s, so too did the energy security goals of *cleanliness and sustainability*, albeit with less initial urgency. As the effects of economic growth and expanding energy use came to be increasingly evident, evolving social, economic, and political values played a significant role in shaping a commitment to protecting the natural environment. Senator Gaylord Nelson, who came up with the idea of Earth Day, was a leader in the effort to galvanize public opinion and turn it into a national movement to advocate for environmental protection. As Senator Nelson stated at an address on the first Earth Day in 1970, "Our goal is not just an environment of clean air and water and scenic beauty. The objective is an environment of decency, quality and mutual respect for all other human beings and all living creatures."[10] The results of these converging concerns about pollution and energy security were both a political and social movement that embraced environmental stewardship, and a rapid change in public law that began to focus more heavily on energy security as an issue of environmental protection and trans-generational sustainability. These goals reflected the growing value attached to environmental stewardship, mitigating the impacts associated with energy production and use, and assuring the development of future supplies that would have fewer adverse environmental and public health impacts. The harm that had already been done to air and water quality resulted in legislation meant to reverse the damage and limit future negative impacts. The *National Environmental Policy Act (NEPA) of 1969*, the subsequent creation of the Environmental Protection Agency in 1970, the *Clean Air Act of 1970*, the *Clean Water Act of 1972*, and a host of water and air regulations all played a key role in not only mitigating the impacts of energy discovery, development and usage, but in redefining the meaning of energy security in the US.

NEPA, for example, established the idea that the environmental impacts of federal government activities must be given full consideration in agency-level planning and decision-making. Moreover, the requirement to produce an environmental impact statement reaches well beyond government activities and into the private sector. When the Bureau of Land Management leases federal land for coal mining, or when the EPA approves a permit for oil and gas exploration, stating that water quality in the area will not be harmed, NEPA is designed to ensure that private sector activities are in compliance with federal law. In the case of the *Clean Air Act*, this statute provides one of the most important ways in which the electric power sector is regulated. For example, the New Source Review permitting program, the Cross State Air Pollution Rule and the Mercury and Air Toxic Standards require that new and existing power plants take systematic measures derived from the best applicable science to protect air quality. To that end, the EPA requires pollution control equipment to capture most emissions of nitrogen oxide, sulfur dioxide, particulate matter, mercury and other toxic and environmental harmful emissions. The EPA also established new rules in 2015 to require the capture of the greenhouse gas carbon dioxide from both new and existing power plants, though these measures, particularly the *Clean Power Plan*, which is designed to cut emissions in existing plants by an average of 30%, are expected to remain contested for some time.

Larger than environmental considerations, and intertwined with them, is the evolving value of *sustainability*, a concept that embodies a concern for generational equity—for future human, economic, social and environmental welfare affecting yet unborn people. Its manifestation in energy policy has changed over time, especially as concern over global climate change has grown. One way this value has been put into effect is to assure future supplies of energy that are cleaner than fossil fuels, and potentially more sustainable as to their sources. Thus, the federal government and a number of states have adopted public policies to incentivize renewable energy sources. Wind and solar power are two of the most popular and fastest growing such alternatives, as they use "fuel" that doesn't pollute and doesn't run out (though these technologies are not without their own adverse environmental impacts). In addition, biomass, geothermal, fuel cells, small hydro (which has less impact than the large dams built in an earlier era), along with wave and tidal power are all also considered renewable energy resources. In the 1970s and 1980s, federal policy began to offer support through tax incentives, most notably the investment tax credit (ITC) and the production tax credit (PTC). The ITC for renewables, begun in 1978 with the *Energy Tax Act*, and greatly expanded with the *Energy Policy Act of 2005*, provides a tax credit in the amount of 30% of the installed cost of a qualified renewable energy system. The PTC, which primarily targets utility-scale projects, began with the *Energy Policy Act of 1992* and has provided a credit of 2.3 cents per kilowatt-hour for eligible sources.[11] These incentives have changed in amount over the years (the ITC went from 10% to 30% in 2005, and the PTC is indexed for inflation), and the PTC has been allowed to lapse several times, including at the end of 2014. These incentives helped to spur development of the renewable energy sector. As concern over climate change has grown, so too have these markets grown rapidly both in the US and abroad. For example, in 2014 wind power accounted for roughly 4.5% of all the electricity consumed in the United States; the level of installed solar PV capacity grew by 76% in 2012, an additional 40% in 2013, and then 30% in 2014.[12] By 2013, the price of solar panels had dropped 99% since the 1970s, from roughly 75 dollars per watt to 75 cents per watt.[13] These same market dynamics have arisen in other industrialized countries as sustainability promotion has taken on global proportions.

The relevance of sustainability as a value embodied in domestic action has had its international counterpart in good part due to the efforts of the United Nations and other international organizations. Published in 1972, *The Limits to Growth* report provided insight into the rate of use of global resources and the implications of this fact for future generations.[14] At projected rates of growth, it was predicted that widespread economic and social collapse would occur sometime in the 21st century. The Brundtland Commission Report, *Our Common Future*, released in 1987, echoed the concerns addressed in *The Limits to Growth*.[15] These broadly distributed and widely read reports represented potentially worst-case scenarios, and while they faced scrutiny and well-deserved criticism, they nonetheless reflected the growing global concern over continuing unsustainable practices. The UN Conference on Environment and Development, held in Rio de Janeiro in 1992, served as an important point in global commitment to environmental protection and sustainability, and led to the adoption of the *Kyoto Protocol* to curb carbon dioxide emissions. The subsequent establishment and work of the International Panel on Climate Change, which has issued several reports detailing the evidence for and documentation of the adverse impacts of climate change, further demonstrates this growth of sustainability as a global value. These developments reflect a growing sense that some types of economic and social changes need to be established on a global basis in order to promote the objectives of human/social welfare and environmental quality that comprise the concept of sustainability. A key part of this global agenda for the future includes reductions in the use of non-renewable sources of energy and the promotion of cleaner renewable energy alternatives so that the global "budget" for carbon emissions will not be exceeded. Energy security viewed from a sustainability perspective is to be promoted through changes in energy portfolios—managed not solely with respect to market exchange considerations, but managed in accord with national policy guidelines and international agreements in which sustainability is a meaningful goal.

The United States has generally been supportive of these global developments, and has sometimes been a leader, but it is not always in sync with the international community, particularly with regard to climate change. For example, with respect to global climate change, the United States government has continually faced criticism from environmentalists both at home and abroad for not taking stronger action. President George H.W. Bush attended the 1992 Rio conference and found the US heavily criticized for not making a more significant commitment to initial international efforts to promote sustainability and environmental quality, seen by other nations as being inextricably linked to issues of energy security.[16] Five years later, the United States signed, but did not ratify, the Kyoto Protocol which was signed and then subsequently ratified by 191 countries, and went into effect in 2005. The lack of any further concerted global action on climate change continues to be considered, at least in part, a major failure of US international leadership and domestic policy by the environmental community.

Energy and National Security

The pursuit of energy security should not conflict too greatly with other major values, and national security is one such major value. Since the 1973 oil embargo there has been an ongoing concern (one that has ebbed and flowed) about the national security implications of America's patterns of energy use. These patterns will be discussed in detail later, but in short they have been most prevalent with respect to oil supplies from the politically volatile Middle East, prompting threats and action against those who might threaten oil exports, and the sizable presence of the US military in the region since 1990. At the same time, the national security implications of energy also include the issue of nuclear power and nuclear weapons proliferation; the ability of energy-rich states to gain global power and influence; the growing energy demands of China and other rapidly developing countries, a development that is increasing the worldwide competition for energy resources; the potential impacts of global climate change on conflict between and within states; and the uncertain impacts of America's growing oil and natural gas production (the oil price drop in 2014–15 has had impacts as varied as slowing US oil production in North Dakota and Texas, creating turmoil among OPEC countries regarding their own proposed output, and causing a drop in the value of the Russian ruble and Norwegian krone).

The impacts of America's energy needs on its domestic and foreign policies are in fact many. With regard to foreign policy, first and foremost is the massive military presence that the United States has maintained in the Middle East since 1990. The United States dispatched a force of more than half a million servicemen and women in the wake of the Iraqi invasion of Kuwait. It then maintained a military presence in the no-fly zone of northern Iraq for more than a decade, while stationing US troops in Saudi Arabia and Kuwait, and later in Bahrain, Qatar and other countries in the Persian Gulf. After 9/11, the war in Iraq further expanded America's military presence in the region. This war, along with the war in Afghanistan, has involved putting American troops in harm's way for more than a decade. Combined, both conflicts have cost hundreds of billions of dollars, and have demonstrated the clear limitations of American power and influence in the region. While these wars were not entered into with the major goals being the protection of oil resources and energy markets (and in Afghanistan, such objectives have not been part of the mission at all), the role of energy resources has been central to both of these conflicts. Iraq's invasion of Kuwait prompted the American response to ensure that Iraq would not control an even larger proportion of the region's oil exports. The subsequent American military presence in Saudi Arabia helped to fuel al-Qaeda and stimulate the attacks by this organization on the United States, including those carried out on September 11, 2001 against targets in New York City and Washington, DC, attacks which led directly to the wars in Afghanistan and Iraq. The withdrawal of troops from Iraq and the ongoing effort to end US military involvement in Afghanistan, combined with growing oil and gas production in the United States, have taken some of the political urgency out of the national security threats emanating from America's energy use. Still, the policy goal has been a consistent one. Throughout the post-World War II period, the United States has maintained an expensive long-standing commitment to maintaining the flow of oil as a national security interest. This interest was made quite explicit in the Carter Doctrine, a response to the Soviet invasion of Afghanistan, in which the United States threatened military action against any nation that jeopardized the continued supply of oil to the United States from the Middle East. These commitments over the past decades reflect the reality that even though the United States has pursued multiple interests in the Middle East, the devotion of blood and treasure (or the threat to do so) would not be the same if there were no oil in the region.

The United States has responded to these external circumstances with domestic action to diversify energy supplies and make the energy supply chain more robust, not only to increase reliability and sustainability but also to enhance national security. It is no accident that major energy legislation has passed either in the wake of or in the midst of US military actions in the Middle East. The *Energy Policy Act of 1992*, the *Energy Policy Act of 2005*, and the *Energy Independence and Security Act of 2007* were all, in part, responses to the growing realization that oil dependence had adversely (and increasingly) impacted the security of the United States. So too were the energy provisions of the *American Recovery and Reinvestment Act of 2009*. All of these Congressional actions, along with many others passed in this time period, have sought to encourage the expanded production and use of alternative energy (particularly biofuels, nuclear, solar and wind power) and fossil fuels (clean coal, unconventional oil and gas) alike. They have also addressed the efficiency and reliability of the electric power grid, informed by an understanding that the grid is highly vulnerable to major disruption (by both deliberate attack and unexpected natural disasters or technical failures).

To that end, the United States has taken myriad actions, including:

- Mandating the production and use of biofuels, with a goal of producing 36 billion gallons by 2022.
- Appropriating billions of dollars for the development of carbon capture and sequestration technologies, also known as "clean coal."
- Offering loan guarantees for the development of nuclear power plants (though in the wake of the nuclear energy crisis in Japan, nuclear energy is unlikely to witness a major resurgence in the short term).

- Extending tax credits for renewable energy technologies, leading to widespread small scale and large scale installations of solar and wind power in the United States.
- Providing permits and royalty payment relief for deepwater drilling in the Gulf of Mexico.
- Deregulating markets in natural gas and electricity to increase supplies, and to make energy markets more efficient and responsive.
- Exempting oil and gas production from certain provisions of the *Safe Drinking Water Act*, the *Clean Water Act* and the *Clean Air Act* to allow for the speedy development of unconventional oil and gas resources through the use of "fracking" and horizontal drilling.
- Investing billions of dollars in "smart grid" demonstration projects designed to modernize and better secure the nation's electric power grid.
- Creating an Advanced Research Projects Agency for Energy (ARPA-E) to support the development of new and innovative energy technologies in the nation's private, public and non-profit sectors.

The Current Landscape

The evolution of US energy policy has been characterized by the ongoing redefinition of the concept of energy security. All of the federal legislation described above, along with numerous other executive agency actions, has promoted the objective of diminishing the vulnerabilities that have emanated from the country's energy needs and prior policy choices. The result is that today, particularly in the aftermath of 9/11, there has been a substantial convergence of values, one in which concerns about national security, economic prosperity, environmental protection, and long-term sustainability all point to specific energy policy choices. In both the academic literature and the popular press, the achievement of energy security—as it is defined in Chapter 1—is seen as a key element in achieving a variety of national security, economic and environmental objectives. The net result for the American political process is that defense and foreign policy, environmental policy, economic policy, and energy policy have all found considerable common ground in the idea of the advancement of energy security.

In one view, this convergence means that a comprehensive push for renewable energy resources is not only one solution to the energy security dilemma, but *the* solution. This view has been most succinctly expressed by the widely read journalist Thomas Friedman of *The New York Times*, as the idea that "green is the new red, white and blue."[17] With respect to the political landscape of the United States, notwithstanding the contemporary goals of environmental protection and sustainability currently garnering great attention, the broad realm of energy policy nonetheless continues to feature the longstanding values that characterized the past (e.g., abundance, affordability). These long-established goals continue to receive important support in public policy along with the more recently established elements of energy security. There is a strong element of "path dependency" in this continuity—the established patterns and practices of energy production, delivery and consumption cannot be easily or quickly altered in a country of 300 million people. The decisions made in the past regarding energy policy now exert influence over the present, placing limits on the range of policy options available, or at least the timeframe in which change can be achieved. Moreover, the politics of energy in the United States still heavily reflect the strength of the oil and gas industries, which continue to benefit from public policies that support their development. Consequently, policies to promote the production and use of coal, oil and natural gas, in spite of their detrimental environmental impact, remain essential pillars of the US energy policy landscape and the politics associated with energy source development, fossil fuel exploration, resource recovery and patterns of energy use.

Needs and Choices

Values are reflected in policy choices, but not entirely. What the United States government and the public say they want with respect to energy and what actually happens in the country may not always coincide. This is especially true because of the large number of public and private actors that help to shape the energy profile of the country, leading to the kaleidoscopic nature of US energy policy noted above. Perhaps the most glaring example of this apparent disconnect involves the goals of environmental protection and sustainability, in which the gap between stated values and actual energy use seems quite large by any measure. The magnitude and urgency of environmental problems suggested by the scientific community, as well as numerous interest groups and policymakers, coincides with a great focus on climate change and renewable energy present in the media and in contemporary political discourse. This public focus on the values of environmental protection and sustainability suggests that there is or ought to be greater effort to more rapidly modify the current US energy profile toward expanded adoption of sustainable energy sources. One might be led to believe from the popular press—due to the number of stories on green technologies, sustainability, renewable energy, and strategies being employed to combat global warming—that a great change has already been achieved with respect to America's energy usage. Along the same lines, advertising by numerous major corporations, including automobile manufacturers, utilities, oil companies, retail stores, and others, would suggest that the United States has already significantly diminished the use of fossil fuels in its energy portfolio. Any such notions are largely inaccurate, however. While the level of renewable energy usage is greater than ever in the past, it remains but a small proportion of the total amount of energy being used in the United States and worldwide. For example, solar and wind power combined accounted for less than 5% of US electricity generation in 2014.[18]

This particular example indicates that while public policy does indeed reflect values, so too do the long-established patterns of production and consumption of energy resources. Therefore, another important way to document the nature of energy choices in the United States, and thus the values that motivate concrete action, is to investigate carefully patterns of actual use. In other words, while policy choices communicate what is valued, so too do our choices regarding the types and amounts of energy that are actually consumed.

The energy used in the US comes from a variety of sources, including oil, coal, natural gas, nuclear, hydropower, biofuels, solar and wind, and most of these sources are used either in part or entirely to generate electricity. The fuels that are used the most, by far, are oil, gas and coal, which combined account for more than 80% of all energy consumed in the United States.[19] Moreover, the amount of energy the United States uses is truly massive. It is sometimes difficult to fully grasp the level of energy produced and consumed, as it is so very large. The following sections of this chapter look at the two ways in which Americans tend to understand their energy needs, as coming from oil (for transportation mostly, but also industry and heating) and electricity (for almost everything else, though direct use of natural gas is important for industry, cooking and heating). These sources of energy largely run the American economy, and without their continued and growing use, the nation's prosperity would be severely threatened.

Oil

With regard to oil, here are the numbers (there's a lot of information here, but stick with us). The level of US consumption in 2014 was 19 million barrels (bbl) of oil per day.[20] (One barrel of oil is 42 gallons. While oil is not sold in barrels, and has not been sold in barrels for decades, it is still measured and reported in this manner, though in other countries it is also measured and reported in tonnes.) The United States is by far the world's largest consumer, comprising about 21% percent of the world's total usage of 90 million barrels per day. Out of its total consumption, the United States imported in 2014 about 40% of its daily needs (down from 60% in 2005), roughly 7.5 million bbl/day, while the rest (including ethanol and other petroleum products) was produced domestically.[21] The next closest consumer in sheer volume is China, whose oil consumption in 2014 stood at 11 million barrels per day, with 6.5 million barrels per day being imported.[22]

Most of the oil used in the United States, about 70%, is used for transportation, with industrial uses and heating accounting for most of the nation's other oil consumption. The transportation sector is particularly dependent upon oil, as 96% of all transport in the United States relies upon oil.[23] Gasoline, diesel fuel and jet fuel account for almost all of this, which includes 253 million cars, SUVs, pickups, semi trucks, construction/work vehicles, and motorcycles; and about 87,000 flights per day for commercial, private, cargo and military airplanes.[24] And of course, total oil consumption is expected to rise in the years ahead. The US Energy Information Administration estimates that total US oil consumption will rise slightly in the coming years, peaking at 19.7 million bbl per day in 2023, then declining back to 2014 levels by 2040. However, global demand is expected to rise a great deal, up to 119 million bbl per day by 2040, leading to greater price increases and significant pressures upon governments around the world to ensure sufficient energy supplies for their respective countries.[25]

The cost of consuming all this oil is quite large. It is determined by factors such as global supply and demand; the strength or weakness of the US and global economies; OPEC's attempts to control output and price; war and political instability in the Middle East; and acts of terrorism in the Middle East and elsewhere that generate concerns about future oil supplies (the price impact of these concerns is also known as the "fear premium"). The 19 million barrels per day the United States consumed in 2014 would cost $570 million if oil were priced at $30/bbl, the level it fell to in early 2016. This level of consumption amounts to an annual cost of $208 billion. At the peak price of $145/bbl, which was reached in July of 2008, the daily cost of oil consumption in 2014 would be $2.8 billion, which amounts to roughly one trillion dollars per year. In June 2014, West Texas Intermediate (WTI) grade crude oil sold at $100/bbl, which translates into a daily cost of $1.9 billion, and an annual cost of roughly $700 billion. At that cost and levels of usage, the US sends roughly $750 million overseas every day, or $274 billion per year, to pay for its oil needs.[26] By the end of 2015, the price of WTI crude oil had dropped by two-thirds from the previous year's high, providing a massive financial benefit to consumers. (See Table 2.1.)

TABLE 2.1 US Oil Consumption and Costs

Total US/Global Consumption – 2014	19 million/91 million bbl/day
US as % of Global Demand	21%
Total Projected US/Global Consumption – 2040	19 million/119 million bbl/day
US as Projected % of Global Demand – 2040	16%
US Oil Imports – 2014	7.5 million bbl/day
Imports as % of US Consumption – 2014	40%
% of All Oil Used for Transportation	70%
% of Transportation Using Oil	96%
Total Daily/Annual Cost of 19 million bbl/day at $30/bbl	$570 million/$208 billion
Total Daily/Annual Cost of 19 million bbl/day at $102/bbl	$1.9 billion/$700 billion
Total Daily/Annual Cost of 19 million bbl/day at $145/bbl	$2.8 billion/$1 trillion
Total Daily/Annual Cost of Imports (7.5 million bbl/day) at $100/bbl	$750 million/$274 billion

Source: US Energy Information Administration.

Electricity

That's a snapshot of American oil consumption; now let's look at electricity. While the unit of measurement for oil is the barrel, for electricity it is the watt, named after James Watt, the Scottish engineer and inventor who lived in the 1700s and early 1800s. One watt will not get you very far—a typical light bulb uses 100 watts. Ten such light bulbs will require one thousand watts, which are called a kilowatt (kW), and a thousand kilowatts gets you a megawatt (MW). One thousand megawatts are called a gigawatt (GW). When measuring usage of electricity in homes and businesses, the preferred unit is the kilowatt; when looking at power plant production, megawatts are usually employed, and when looking at national or global consumption, the gigawatt is often a useful scale.

There are two ways in which these terms are employed. One usage refers to the capacity of either: 1) a power plant or smaller system that generates electricity; or 2) a product/system/building that uses the electricity. To that end, there are 100-watt light bulbs, 900-megawatt power plants, or 500-gigawatts of total generation capacity in a given country. Capacity refers to the total ability of a power plant or a product to produce or use a particular amount of electricity. With respect to production, simply because a power plant has the capacity to produce 900 MW at any given time, it doesn't mean that the power plant will actually be generating that much electricity at all times. This is because power plants throughout the country contribute to the grid in different ways. Some plants serve as baseload, and they are operating near full capacity most of the time. However, because electricity demand fluctuates during the day and over the course of a year, there are plants that serve intermediate and peak load. Intermediate load plants may only be used for part of the day or week, and those that serve peak load may only generate electricity at their full capacity a small part of the time, for a few hours a day, or maybe only on the coldest or hottest days of the year.

A second usage of these terms involves the output or consumption of electricity that is actually generated over a particular period of time. For this, the term used is kilowatt-hour (kWh), which is the use of 1000 watts of electricity for one hour (using 10 of those 100-watt light bulbs in your house for an hour). At higher levels of generation and consumption, the larger figures—megawatt-hour (MWh) and gigawatt-hour (GWh)—are used. To that end, a home may use 500 kWh a month, or a mid-sized business might use 800 MWh per month.

With regard to looking at levels of capacity and generation in the United States, the numbers are, unsurprisingly, quite large. Total US generation in 2014 was 4.1 million GWh (or 4.1 trillion kWh). This level is, of course, expected to rise in the future. The US Energy Information Administration projects that this figure will increase to 5.1 million GWh of generation by 2040.[27] To place this number in perspective, the average home in the United States uses about 11,000 kilowatt-hours per year, or 0.011 GWh.[28] (Another way to look at this comparison is that the total national usage per year is 4.1 trillion kWh, versus 11,000 kWh per home.) By way of comparison, in 2013 China's level of electricity generation was 4.9 million GWh (China surpassed the US in 2011, and is expected to use more than twice the electricity of the US by 2040).[29] At the national level, China's rapid growth has led to great increases in electricity consumption, and its aggregate demand has become comparable to that present in the United States. However, the intensity of use is still widely divergent, as the United States has less than one-quarter of the population of China. Each American, on average, uses about five times the amount of electricity per year as the average person in China. These two countries alone account for a very large portion of global electricity generation, roughly 40%. In 2013, the world's total generation was 21.8 million GWh (or 21.8 trillion kWh).[30]

The capacity required to generate all this electricity comes from a variety of power plants, using a variety of fuels to power their respective generators. (Electricity is termed a "secondary" source of energy for this reason, as it requires other energy sources as input.) The total US generation capacity, like total consumption, is also very large. (The numbers characterizing capacity are far smaller than the numbers associated with consumption in kilowatt-hours. This is because a 1 MW plant can generate, in theory, 8,760 MWh over the course of a year, as there are 8,760 hours in a year. Power plants, however, do not operate at full capacity all the time.) In 2013, the United States had 1,081 gigawatts (GW) of generating capacity nationwide (by way of comparison, in 2013 China had a total installed electricity generating capacity of 1,162 GW, up from 525 GW in 2005).[31] The United States is expected to add new capacity in the coming decades, with the US Energy Information Administration estimating that US generation capacity will rise to 1,293 GW by 2040.[32] This generation capacity comes almost entirely from power plants distributed around the country, but there are also a number of on-site generation systems that are installed in buildings, factories, college campuses, houses and remote areas to provide electricity at these particular locations.

The most common fuel used to generate electricity in the US is coal. The US has abundant supplies that are readily accessible at an affordable price, and as a result it was used in 2014 to generate 39% of the electricity used—1.6 million GWh. After coal, the next most common fuel is natural gas, which accounted in 2014 for 27% of all US generation, or 1.1 million GWh. This is followed by nuclear power, at 19% or 797,067 GWh, and large scale hydroelectric, with 6.5% or 258,749 GWh generated.[33]

With respect to renewable energy resources beyond large-scale hydroelectric power, there are solar and wind power, which are currently the most popular and most widely used renewable energy technologies worldwide, but this category also includes geothermal power, hydrokinetic power, biomass/biofuels, and fuel cells, as well as a whole host of new, innovative technologies that are as of now unproven for large-scale use (e.g., bacteria-powered batteries, airborne wind turbines tethered to the ground to take advantage of higher-altitude wind resources). Wind and solar electric power, to take the two most popular and mature technologies, account for very little of the US total, though wind power use has grown quickly since the late 1990s. In terms of generation, wind produced 181,791 GWh in 2014, which was 4.5% of the total used, while solar electric (photovoltaics [PV] and concentrating solar power [CSP]) accounted for 18,321 GWh, about 0.4% of the total.[34] In terms of generation capacity, there is a total of more than 66,000 MW of wind power installed in the United States in 2015, and just over 22,000 MW of solar capacity (out of a total of approximately one million MW).[35] In other words, the actual electricity generation of wind and solar is five percent of the nation's total, while their generation capacity is eight percent of the total. The reason for the divergence in percentages of generation vs. capacity is that the output per unit for wind and solar is comparatively less than coal and natural gas. Compared to coal and natural gas, wind and solar tend to have lower levels of efficiency and energy intensity, along with lower capacity factors (capacity factor refers to the proportion of time during which electricity can be generated at a power plant; the wind isn't always blowing and the sun isn't always shining, but coal and natural gas can always be burning). Again, we have provided a lot of numbers to consider. Table 2.2 below provides them in a more readily digestible form.

The cost to consumers for purchasing all 4.1 trillion kilowatt-hours of this electricity comes to hundreds of billions of dollars every year. Costs to producers

TABLE 2.2 Electricity Generation, Capacity and Sources

Total US Generation 2014	*4.1 million GWh*
Total Projected US Generation 2040	5.1 million GWh
Total US Capacity 2014	1,081 GW
Total Projected US Capacity 2040	1,293 GW
Total China Generation 2013	4.9 million GWh
Total China Capacity 2013	1,132 GW
Total Global Generation 2013	21.8 million GWh
Total Projected Global Generation 2040	39 million GWh
Annual Generation of Average US Home	11,000 kWh/year
Electricity Generated from Coal 2014 / % of Total	1.6 million GWh / 39%
Electricity Generated from Natural Gas 2014 / % of Total	1.1 million GWh / 27%
Electricity Generated from Nuclear 2014 / % of Total	797,067 GWh / 19%
Electricity Generated from Large Hydropower 2014 / % of Total	258,749 GWh / 6.5%
Electricity Generated from Wind 2014 / % of Total	181,791 GWh / 4.5%
Electricity Generated from Solar 2014 / % of Total	18,321 GWh / 0.4%

Source: US Energy Information Administration.

vary among fuel types, but in general the retail price of electricity paid by consumers reflects a blend of these sources, as utilities do not rely on a single type of generation source to supply their customers. In 2014 the price that residential customers paid for electricity ranged from about 8–18 cents per kWh across the country, with an average of 12.5 cents/kWh. Commercial and industrial customers tend to pay lower prices; in 2014 their average costs were 10.8 and 7.0 cents/kWh, respectively.[36] The US Department of Energy (DOE) collects these data on electricity prices throughout the country, to see what traders and consumers are actually paying in wholesale and retail markets. DOE also analyzes the anticipated costs to producers of generating electricity. One way that these costs can be compared on an equal basis is to look at the *levelized cost of electricity* (LCOE). This is a standard practice in the utility industry that takes into account the expected electricity generation and the average cost of that generation over the lifetime of a project. This figure includes initial investments in capital equipment, labor, facility construction, plant siting, and regulatory permitting, along with ongoing fuel costs, staffing, debt servicing, operations and maintenance. This figure does not reflect the actual price of electricity on the market, but rather the price required to cover the project costs. With regard to particular fuels, coal and natural gas are both relatively cheap and plentiful in the United States, and building a power plant to burn these fuels and generate power is relatively inexpensive compared to the alternatives, though new regulations regarding carbon and other emissions are expected to increase the cost of coal-fired electricity in the future. As a result, many coal plants are being retired, and few new ones are being built or planned. Producing electricity from large hydroelectric dams, which were built decades ago, is also relatively inexpensive. The fuel—flowing water—is free, so the electricity rates charged to customers are determined principally by the cost of maintaining the dam and power plant facilities. Even nuclear power, which represents a much more complex undertaking, still manages to generate electricity at a price that is comparable to coal and natural gas. By contrast, solar and wind power have tended to be more expensive than either coal or natural gas in the past, especially solar power. Wind power is increasingly approaching "grid parity" with coal and natural gas, and solar power is continuing to make great gains in cost reduction, with utilty-scale projects approaching grid parity. While the fuels, sunshine and wind, are free, the technology to capture these resources and turn them into usable energy has historically been more costly compared to conventional power plants. The calculation of costs, especially those paid by the consumer, must also be understood in light of the fact that the United States provides large direct and indirect subsidies to facilitate the production of all forms of energy, including these renewable sources.

The US Energy Information Administration periodically projects levelized costs of electricity for different sources. Their estimates, for systems coming into service in 2020, are that electricity from natural gas will be the least expensive, at 7.2 cents/kWh. Wind is expected at 7.4 cents/kWh, with conventional coal and nuclear at 9.5 cents, solar PV at 12.5 cents, and coal plants using carbon capture and storage technologies averaging 14.4 cents/kWh.[37]

There is another way to look at US energy sources and how they are consumed, and this type of measurement is provided in Figure 2.1 below, which characterizes energy in terms of BTUs (British Thermal Units). In the diagram, which displays 2013 energy data, the total annual energy usage in the United States was 97.4 quadrillion BTUs, or "quads" (this total includes oil and other fuels in addition to electricity generation).[38]

This graph succinctly illustrates all US energy production and usage, and clearly shows the relative prevalence of different fuel sources, as well as their uses and how efficient they are (the level of energy lost is indeed staggering). The visual representation reinforces the abundantly clear point that coal, oil and natural gas power the United States of America, accounting for 82% of all energy used in the US. By contrast, just over 9 quads came from solar, wind, geothermal, hydro and biomass power combined, which is 9.4% of the total, and solar and wind accounted for only 1.9 quads of energy (up from 0.156 quads in 2005[39]), less than two percent of the total. As was noted above, despite the prevalence of news stories, websites and advertisements proclaiming the increased use of renewable energy sources, the United States and the world have only just begun to take the first steps toward the widespread use of renewable energy sources in their respective energy portfolios. Another key piece of information that the graph clearly depicts is the massive opportunity for greater energy efficiency. Well over half of all the energy produced from all sources is wasted or "rejected energy," largely in the form of heat, as efficiency losses occur in the conversion from fuel to the intended uses.

The Promise of Policy

The choices made in the United States—in terms of both energy policy and energy consumption as they have been characterized above—demonstrate the energy security dilemma. Conflict arises, and is seemingly inevitable, among the interplay of articulating values, establishing policy, and producing and consuming energy. Varied approaches to energy security are shaped by the values of energy policy stakeholders, such as governmental actors and institutions, businesses, and citizens. The relationships between and among each of these stakeholders are neither uniform nor consistent. At times, stakeholder interests converge; at other times their interests are highly divergent. This makes energy security a highly complex policy area involving numerous broad domestic and global policy dynamics associated with environmental protection policy, renewable energy source incentives, fossil energy source development, changing energy demand profiles, regional geopolitics, global demography, and international climate change phenomena all in some state of flux.

Over the last century, public policy has focused on energy security by means of pursuing its "public good" aspects. During this time, the list of public goods has

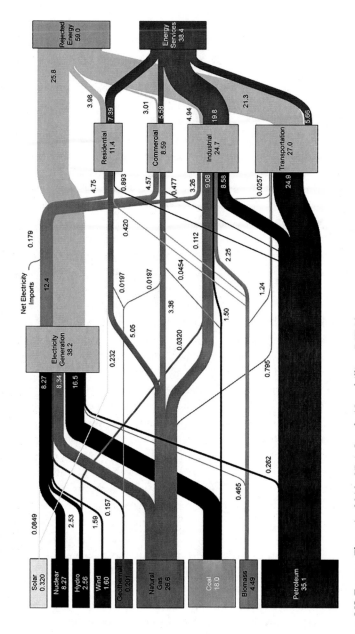

FIGURE 2.1 US Energy Flows in 2013: 97.4 Quads (Quadrillion BTUs)
Source: Lawrence Livermore National Laboratory, 2014.

grown as values have evolved, and unintended outcomes in the pursuit of energy security have resembled the "tragedy of the commons," prompting new policy intended to counter these "tragedies."[40] Achieving abundance and affordability made for a big energy infrastructure, but the apparent success was short-lived as the complex network that was developed created liabilities in maintaining a reliable supply chain. Reliability, however, was not enough, as environmental and national security concerns required attention too, ultimately leading to public policy embracing environmental protection, renewables, and biofuel development, among other initiatives.

The development of alternative energy sources, especially from renewables such as solar, wind, geothermal, biofuels and other resources, currently comprises a key element of energy policy. Such support is based on the idea that over time, but not too much time, the United States, along with the rest of the world, is going to have to produce and use energy differently—with substantially greater efficiency, with more reliance on alternative energy sources, with fewer environmental pollutants—to avoid or diminish the costs and risks associated with current policies and practices. Alternative energy resources such as wind and solar are expected to mitigate or prevent the anticipated problems and crises that seem to be approaching as a result of continued reliance on oil, coal, and natural gas. Moreover, these technologies have captured the greatest attention in the United States, and perhaps throughout the world—far more so than others—because the economic, security and environmental benefits that they potentially offer are both significant and enticing. Many of these technologies hold great promise for the future, and the policy support and investments they are attracting seem warranted because of all the energy security benefits they offer. Increasing the production and use of these alternative technologies can potentially provide clean, reliable, affordable and abundant energy supplies for both electricity generation and transportation, and would go a rather long way toward meeting many of the nation's multiple public policy objectives.

These objectives are all related, as abundant supplies will keep prices down in the long term, reliable supplies will diminish price spikes and volatility, clean and sustainable supplies (which are clean not only with regard to outputs, but with respect to their inputs as well) can provide abundance and thereby affordability, and diverse supplies can help achieve affordability, abundance and reliability, while reducing national security vulnerabilities. This circumstance represents a "virtuous circle" as opposed to an inescapable dilemma, and it represents the ultimate promise of policy. Unfortunately, this virtuous circle thus far has proven to be remarkably difficult to achieve in practice.

It may be the case that a new technology will be developed that can provide massive amounts of energy in a technically feasible, clean, efficient, cost-effective manner. Current alternative energy technologies are getting better and new ones are under development all the time. Solar panels have greater efficiency than ever before, converting a higher percentage of the solar radiation they capture into electricity. Companies, national labs and entrepreneurs are also working on emerging technologies such as thin-film solar, algae-based biofuels, ocean turbines, batteries and a host of other possibilities. However, it is unrealistic and irresponsible to expect that a solution will emerge from a technological breakthrough that can change the entire energy landscape in a single lifetime.

Maybe there are financial or market-based mechanisms that can change how energy markets are structured, and it is possible that these will offer a solution to making alternative energy sources more widespread. One example is the widespread use of the power purchase agreement (PPA) in the market for solar PV generation systems. For such projects, their high up-front capital costs have been a significant barrier to sales and market development. In addition, there are organizations, such as schools, non-profit organizations, government agencies and small businesses that may not have a tax liability to take advantage of federal and state tax credits. The PPA offers a way around these problems, as it involves three parties: the electricity purchaser or end-user, the vendor, and a financier. The financier owns the system, which is located on an end-user's property (e.g., on the rooftop or an adjacent plot of open space), and then sells the electricity to the end-user. The financier pays for the system and maintains legal ownership of it for the life of the contract, which is usually 15–20 years. The financier also receives all federal and state incentives, passing those savings along to the purchaser, who agrees to purchase all of the electricity generated by the system at an arranged price.[41]

Another example of a market-based approach to energy security was provided by the Israeli company Better Place, a corporation that sought to structure an electric car market on the model of the cell phone market. In this business model, the consumer would purchase a product (a car), and subscribe to a monthly service to use it. The service consisted of a network of battery switching stations spread throughout the country. Drivers would exchange batteries that used up their charge for recharged batteries in the same time it would take to fill up a tank with gas. This network of battery stations allowed drivers to pay for the cost of fuel (electricity) over the lifetime of the car, which is by-and-large how they would pay for gasoline. Thus, the initial cost of the car to consumers could possibly be reduced, making electric cars more attractive to consumers than gasoline-powered vehicles. The venture failed and went bankrupt after only a year in business, suggesting a wide gap existed between the idea and its actual practice. Still, the PPA and Better Place both represent business arrangements to allow for added production and consumption of energy alternatives. However, relying too heavily on the market for an energy solution, like relying too heavily on technology development, is an unwise strategy.

This is where public policy plays a key role, as another realm that can have an impact on energy security, on the market, and on what kinds of energy are produced and used in a society. In fact, it represents a realm where progress could potentially be made quite rapidly. Technology breakthroughs cannot be planned, and new business arrangements can only do so much to reshape the market in light of current technologies and their costs. In the public policy realm, however, there is ongoing opportunity, and there appears to be value that has not been captured. There are, of course, numerous examples of energy policy measures that have created or worsened problems.[42] The most likely way to start the process of meeting multiple energy security objectives—ensuring supplies, avoiding security risks, diminishing economic vulnerability, and protecting the environment—is to make policy work well. This can involve supporting and incentivizing markets for new and expanded energy supplies via incentives and mandates; limiting the potential for international conflict; cleaning up and protecting the environment by limiting environmental stresses and the human impacts they can cause.

With respect to energy policy, analysis tends to come in three different forms:

1. *Supply expansion.* This approach seeks to determine the extent to which policy measures affect the expansion of different forms of energy supplies. Analyzing policy through this lens tends to be based upon an understanding that energy needs are growing, and that this demand for energy needs to be met. The problem according to this analytical framework is not energy shortages—there are, after all, a wide variety of energy resources available— but of converting these resources into usable energy. The focus of this framework is on the barriers that exist to such a goal, be they technical, policy (statutory or regulatory), economic, environmental, geographical or political, and how to overcome them through policy solutions (financing, tax incentives, direct appropriations, demonstration projects, statutory or regulatory mandates). Therefore, the key to determining the value of a policy is to measure it against the realization of this goal.

2. *Demand management and reduction.* This approach takes the opposite tack, determining the extent to which policy measures affect the reduced use of energy—in other words, how well policy promotes and achieves efficiency and conservation. Analyzing policy through this lens acknowledges that energy needs are growing, and that new supplies are indeed necessary, but it also suggests that new supplies are likely not to be sufficient. Because so much energy is wasted through overconsumption and inefficiency, one of the best ways to become more energy secure is to reduce demand and enhance efficiency. The policy focus here has been on incentives (tax credits and rebates for energy efficient equipment), consumer education (energy rating information on appliances), technological development (smart grid demand management hardware and software), mandates (CAFE standards, building codes), and reducing barriers to lifestyle changes that would result in reduced energy usage (city planning that would encourage walking and biking over driving). This type of analysis seeks to determine the value of policy by measuring it against the realization of this demand-centered goal.

3.   *Cost Analysis.* An approach based on the analysis of costs reflects an understanding that the best energy resources (and this includes efficiency and conservation) are those that minimize the costs associated with their production and use. Costs can be defined narrowly, in terms of financial costs to consumers (the price of a kilowatt-hour of electricity or a gallon of gasoline), and to tax-payers (government appropriations and tax expenditures). They can also be defined more broadly, in terms of environmental, national security, and social costs. This can include examining the health and environmental impacts of coal (mining, impoundment of tailings, transportation, and burning it in power plants). It can also include the cost of military operations incurred in the protection of oil and oil markets (or some portion of those costs), such as those in Iraq or the larger Persian Gulf region. Such an approach at its best seeks to capture not only the expected costs, but also the unintended costs of energy resources and the governmental policies involving them.

It is a combination of the first and third of these approaches that grounds the analysis here. As successive presidents, Congresses, Governors, legislatures, and interest groups seek to enact energy policy, their support for various legislation and regulations is meant to achieve a set of desired aims. As implementation of policy proceeds, however, both the expected and unexpected costs of pursuing a given public policy measure begin to manifest themselves, demonstrating the dilemma that is inherent in the pursuit of energy security.

Equipped with our understanding of US energy security and policy analysis, along with the values, choices and needs that inform policy, the following chapters provide a critical analysis of policy adopted in the United States in the pursuit of energy security.

Notes

1 Severine Deneulin and Nicholas Townsend, "Public Goods, Global Public Goods and the Common Good," WeD Working Paper 18, September 2006; and Marc Wuyts, "Deprivation and Public Need," in Macintosh, M., Wuyts, M. and Hewitt, T., *Development Policy and Public Action*, Oxford: Oxford University Press, 1992.
2 See June Sekera, "Rethinking the Definition of 'Public Goods,'" *Real World Economics Review Blog*, July 9, 2014, https://rwer.wordpress.com/2014/07/09/re-thinking-the-definition-of-public-goods/ (accessed January 30, 2015).
3 WTRG Economics, "Oil Price History and Analysis," www.wtrg.com/prices.htm (accessed October 18, 2013).
4 US Department of Energy, "Energy Efficiency and Renewable Energy, Historical Gasoline Prices, 1929–2011," www1.eere.energy.gov/vehiclesandfuels/facts/2012_fotw741.html (accessed October 18, 2013).
5 "The Arab Oil Threat," *The New York Times*, November 23, 1973, p. 34.
6 Center for Energy and Climate Solutions, "Federal Vehicle Standards," www.c2es.org/federal/executive/vehicle-standards (accessed October 18, 2013).
7 United States Department of Energy, Office of Fossil Energy, "SPR Quick Facts and FAQs," http://energy.gov/fe/services/petroleum-reserves/strategic-petroleum-reserve/spr-quick-facts-and-faqs (accessed September 1, 2015).

8 "Using Biofuel Tax Credits to Achieve Energy and Environmental Policy Goals," Congressional Budget Office, July 2010, p. 2; and US Energy Information Administration, "Energy Timelines, Ethanol," www.eia.gov/kids/energy.cfm?page=tl_ethanol (accessed October 18, 2013).

9 "Biofuel Trade Organizations Ask for Extension of Advanced Biofuel Tax Incentives," Renewable Fuels Association, July 17, 2015, www.ethanolrfa.org/news/entry/biofuel-trade-organizations-ask-for-extension-of-advanced-biofuel-tax-/ (accessed September 1, 2015).

10 "Gaylord Nelson and Earth Day," Nelson Institute for Environmental Studies, University of Wisconsin, www.nelsonearthday.net/nelson/earthdayidea.htm (accessed February 3, 2015).

11 Database of State Incentives for Renewable Energy (DSIRE), "Federal Incentives/ Policies for Renewables and Efficiency," www.dsireusa.org/incentives/index.cfm?state=us (accessed October 18, 2013).

12 US Energy Information Administration (EIA), *Monthly Energy Review*, August 2015, p. 95; Solar Energy Industries Association, *U.S. Solar Market Insight Report, 2014 Year in Review*, p. 3.

13 "Pricing Sunshine," *The Economist*, December 28, 2012.

14 Donella Meadows, Jorgen Randers, Dennis L. Meadows, and William W. Behrens, *The Limits to Growth: A Report for the Club of Rome's Project on the Predicament of Mankind*, New York: Universe Books, 1972.

15 World Commission on Environment and Development, *Our Common Future*, Oxford: Oxford University Press, 1987.

16 Christopher A. Simon, *Public Policy: Preferences and Outcomes*, New York: Longman, 2010.

17 Thomas Friedman, "The Power of Green," *New York Times Magazine*, April 15, 2007.

18 EIA, *Monthly Energy Review*, August 2015, p. 107.

19 EIA, *Annual Energy Outlook 2015*, Total Energy Supply, Disposition, and Price Summary.

20 EIA, *Annual Energy Outlook 2015*, Petroleum and other Liquids Supply and Disposition.

21 EIA, *Annual Energy Outlook 2015*, Petroleum and other Liquids Supply and Disposition.

22 EIA, *International Energy Outlook 2014*, World Liquids Consumption.

23 EIA, *Annual Energy Outlook 2015*, Petroleum and other Liquids Supply and Disposition; and James Woolsey, "How to End America's Addiction to Oil," *The Wall Street Journal*, April 15, 2010.

24 US Department of Transportation, "Bureau of Transportation Statistics," www.rita.dot.gov/bts/sites/rita.dot.gov.bts/files/publications/national_transportation_statistics/html/table_01_11.html (accessed June 6, 2014); National Air Traffic Controllers Association, www.natca.org/ and National Oceanic and Atmospheric Administration, http://sos.noaa.gov/Datasets/dataset.php?id=44 (accessed June 6, 2014).

25 EIA, *Annual Energy Outlook 2015*, Petroleum and other Liquids Supply and Disposition; and EIA, *International Energy Outlook 2014*, World Liquids Consumption.

26 These numbers are calculated and rounded by the authors.

27 EIA, *Annual Energy Outlook 2015*, Electricity Supply, Disposition, Supply and Prices.

28 EIA, "Frequently Asked Questions," www.eia.gov/tools/faqs/faq.cfm?id=97&t=3 (accessed June 6, 2014).

29 EIA, *International Energy Outlook 2013*, World Total Net Electricity Generation.

30 EIA, *International Energy Outlook 2013*, World Total Net Electricity Generation.

31 EIA, *International Energy Outlook 2013*, World Total Installed Generation Capacity.

32 EIA, *International Energy Outlook 2013*, World Total Installed Generation Capacity.

33 EIA, *Monthly Energy Review*, April 2015, p. 105.

34 EIA, *Monthly Energy Review*, April 2015, p. 105.

35 American Wind Energy Association (AWEA), "Wind Energy Facts at a Glance," www.awea.org/Resources/Content.aspx?ItemNumber=5059 (accessed June 6, 2014);

Solar Energy Industries Association (SEIA), "Solar Energy Facts: 2013 Year in Review," March 5, 2014.

36 EIA, *Electric Power Monthly*, February 2015.

37 EIA, *Annual Energy Outlook 2015*, Estimated Levelized Cost of Electricity (LCOE) for New Generation Resources, 2020.

38 Lawrence Livermore National Laboratory, "US Energy Flow," https://flowcharts.llnl.gov/ (accessed June 6, 2014).

39 George M. Whitesides and George W. Crabtree, "Don't Forget Long-Term Fundamental Research in Energy," *Science*, 315(5813), February 9, 2007, pp. 796–8.

40 Garrett Hardin, "Tragedy of the Commons," *Science*, 162(3859), 1968, pp. 1243–1248.

41 For further information, see Solar Energy Industries Association, "Solar Power Purchase Agreements," December 20, 2012, www.seia.org/sites/default/files/resources/SolarPPAs_fact%20sheet_FINAL%201.pdf (accessed December 7, 2015).

42 See Peter Grossman, *U.S. Energy Policy and the Pursuit of Failure*, Cambridge: Cambridge University Press, 2013; Michael Graetz, *The End of Energy: The Unmaking of America's Environment, Security and Independence*, Cambridge, MA: MIT Press, 2013.

3

PRIVATE PROPERTY RIGHTS AND THE PUBLIC GOOD

The core issues of energy security arise somewhere near the borderline between pure public policy considerations and the marketplace of free enterprise entrepreneurial activity. Moreover, energy security is often framed by proponents of one or another specific measure as a pressing domestic concern, when in fact it may often also necessitate that significant attention be paid to current and future international relations and existing and developing global market forces alike. Taken from a domestically focused position, one nation's or one person's "energy security" can lead to insecurity for others and other nations. Even when discussion is entirely focused domestically, problems arise that have combined national and international implications. Japan's decision to promote the development of nuclear power on a grand scale in the 1950s, and the large scale development of a national electrical energy security solution for a nation whose consumption of imported oil has to be kept to reasonable levels to maintain a favorable balance of trade had global negative consequences in the form of the Fukushima disaster. The release of radioactivity had, and will continue to have, severe national and even international implications, perhaps raising the risk calculations for the use of nuclear energy sources so much that it will not be a politically viable alternative for addressing global climate change issues for a long time into the future. In another example of domestically focused energy security policies, the United States has inadvertently contributed to driving up global food prices as the adoption of ethanol additive policies led to increased demand for corn and sugar cane feedstocks. Yet, despite these clear examples of the interplay between the national energy security and global energy security realms, so much of the energy security policy debate remains a function of principally domestic-focused concerns. Even overriding international concerns, such as global warming and global food system security, are dealt with through a patchwork quilt of public policy,

heavily weighted down by glacial movement in statutory changes and legal precedent. For example, US agricultural policy and spending are authorized every
five years through legislation commonly referred to as the "Farm Bill." This legislation can potentially affect the issues of climate change and global food security
in important ways by mandating or incentivizing particular land uses and commodity production. At the same time, the legislation offers a lesson in how these
two global concerns are filtered through the lens of domestic farm commodity
groups, state and regional governmental actors, and House and Senate committees and subcommittees. The metaphor of glacial movement seems appropriate
indeed.

Policy Analysis and Energy Security

Energy security is a concept in need of a clear definition, hence the considerable
space afforded to that topic in Chapter 1. A broad array of threats to energy
security exist; some are large threats such as instability within and among the
nations of the Middle East in the wake of the Arab Spring phenomenon, and
some are seemingly small threats such as localized energy systems failures in the
form of power grid breakdowns. These threats both large and small endlessly
complicate the analysis involved in managing energy security. And, of course,
there are emerging threats that one might never have considered of any significance only a decade ago, but which are now broadly viewed as substantial
threats, such as computer hacking of power generators and distribution networks
in order to shut down supply or significantly disrupt the power grid of a particular country or region of the world. What is required to define and properly
study the principal public policy dimensions of energy security, therefore, is
comprehensive and well-considered policy analysis.

Philosophy of governance plays a significant role in determining whether
energy security is primarily driven by public policy, by market-driven choice, or
by some combination of both. While elected officials may discuss the generalities
undergirding these distinctions on the election stump from time to time, systematic
policy analysis is the process by which the role of government in the global
energy marketplace is more precisely defined. Policy analysis identifies the negative
externalities that serve to constrain citizens when such constraint may be required,
and it is the instrument by which Pareto optimal choices are identified—that is,
solutions that maximize social benefit at an efficient price vis-à-vis avoided costs to
individuals affected (in other words, when marginal costs and marginal benefits
are equal [MC=MB]).

Identifying an appropriate role for government first requires us to consider
energy as a "good" in the parlance of economics—the benefits and externalities
of the good as well as its ready and affordable availability. Energy security,
therefore, is not just an issue of product safety, although that has played an historically significant role in the development of the petroleum industry, electrical

power, and nuclear power alike. Security has become for governments primarily a supply issue, one that government cannot leave to the market alone because energy supply is intimately associated with the legal transfer of ownership from the sovereign to the private sector, as well as being connected to the broadly understood goals of government in the modern and post-modern age.

Beyond the direct benefits from the use of energy, government seeks to reduce the often-injurious negative externalities associated with energy development, distribution and use. A good example of a negative externality involves climate-changing greenhouse gases such as carbon dioxide, which power plants emit in the course of generating electricity from coal. Hydroelectric dams provide another prominent example of negative externalities—in this case the unintentional impacts on Pacific salmon runs and biodiversity resulting from large scale impoundments of once free-flowing rivers. While the energy produced is itself a *marketable good*, all risks associated with energy production are viewed through the lens of *public goods*, whereby marginal social costs can be understood in the broadest conceptual sense.

Both of these elements are central in understanding the concept of "energy security," which is really a composite of two concepts that are at times disparate. Energy is a product that is needed in the modern world and is treated largely as a *commodity* traded in vast quantities as billions of barrels and quadrillions of watts. Conversely, the word *security* implies a level of risk as being acceptable or unacceptable. While there will always be some form of useable energy, security and acceptable risk taken together constitute a function of values applied to inevitable tradeoffs. And as we pointed out in Chapter 2, societal values have continually shaped policy preferences about energy and energy security, while changing over time. Nor are these values consistent in terms of issue prioritization or preferred public policy response.

In the post-materialist world described in the work of Ronald Inglehart and his associates, acceptable levels of risk have shrunk considerably since the advent of post-industrial society.[1] In part, this is due to increased knowledge arising from investments made in science and technology and more widespread awareness of the complex nature of interdependencies among the nations of the world. There is now virtually instantaneous information about threats to daily life, to include threats to energy supply and the negative externalities of energy use in nearly any part of the world. For example, a decision made in China to continue its reliance on the large-scale importation of coal to provide electrical power causes great activity in the mines of Montana and Wyoming and stirs potential protests in the coastal cities of Washington and Oregon through which China-bound coal will pass for transshipment. The values of citizens and politicians in the Mountain West of the US are "miles apart" from those of the Pacific Northwest with respect to the assessment of risk associated with global climate change. Increasingly, government is called upon to provide a world that is clean, safe, equitable and sustainable while allowing a maximum of freedom in the marketplace for free enterprise and entrepreneurial vigor to "grow the pie" of the national economy. In the case of the United States, the federal government is the inevitable arbiter of value conflicts and the final point of public policy formation for energy policy. As was noted in Chapter 1, the "energy policy" of the nation is much more of a patchwork quilt of conflicting initiatives and declarations of noble intent than a systematic framework for making ongoing critical tradeoffs. This is the case in part because energy security issues are complex and ever-changing, but it is also the case because of the supremacy of private property rights in our constitutional heritage and how that private property rights-based framing of the business of energy development and marketing has developed in our country.

Property Rights and Energy

The legal term property rights quite naturally brings to mind the concept of "real" property. The land owned or the tangible goods possessed by an individual or corporation are seen as solid immutable examples of property. The designation of a property right to the solid stationary energy sources, namely coal, is easily accomplished through land surveys. Other forms of energy, however, do not match up well with our conventional conceptions of property. For example, natural gas and petroleum are capable of migration from one location to another at considerable distance. In other words, gas or petroleum supplies that exist under one parcel of land may migrate to other locations once drilling and resource extraction begins to take place.

Property rights tied to these migratory goods are governed by the legal doctrine of "rule of capture." A "person who owns the surface may dig therein, and apply all that is there found to his own purposes, at his free will and pleasure; and if, in the exercise of such right, he intercepts or drains off … his neighbor's well, this inconvenience to his neighbor falls within the description of *damnum absque injuria*, which cannot become the ground of an action."[2] Very much like catching wild prey, the rule of capture means that those in pursuit must act quickly and with cleverness in order to get the goods! Those who take things slowly and methodically are likely to lose out on the benefits to be had. One way to overcome the problems associated with this legal doctrine is for governmental agencies—in many cases, already tasked with setting withdrawal limits and establishing well-spacing in the interest of conservation and conflict avoidance—and individuals in pursuit of migratory resources to agree that rational resource management, principally through limitations on access, is to the mutual benefit of all parties in the long run and to suspend the rule of capture.[3]

While US state law limits were imposed by a variety of state agencies regulating subsurface property rights on private lands, the *General Mining Law of 1872* created a virtual *laissez-faire* atmosphere for mineral resource development on the nation's public lands.[4] The rush to claim resources on these lands was so swift that it required Presidential Executive Action to set aside land and mineral resources for national strategic purposes. The *Mineral Leasing Act of 1920* led to the establishment of a more coherent approach to claim staking and leasing to protect a natural national resource viewed as a public good; up until that time property rights were established via the *1872 General Mining Law* and the *Oil Placer Act of 1897*.[5] The *Mineral Leasing Act* and its many revisions define property rights associated with mining claims and establish critically important limitations to mineral leases. For example, the law sets aside fossil energy reserves for strategic purposes and for environmental protection. The law also limits the time duration of leases, causing claim holders to move expeditiously in resource withdrawals. Through administrative oversight, the law requires regular evidence that leased lands are indeed being developed for the purposes of fossil energy extraction.

The *Mineral Leasing Act* is administered by the Bureau of Land Management within the Department of Interior and the statute covers both onshore and offshore leases. As of 2012, the onshore oil and gas leases on public lands numbered 48,699 and covered 37.8 million acres. Over 92,000 producible and serviceable well holes were in operation at that time. Offshore leases covered 35.8 million acres of which 10.1 million acres were active leases. The majority of the offshore leases are located in the Gulf of Mexico.[6] In March, 2013, the BLM announced an August, 2013 auction of "all leased [a]creage in the Western Gulf of Mexico, an area in excess of 21 million acres featuring an estimated production of 116 to 200 million barrels of oil and 538 to 938 billion cubic feet of natural gas."[7]

Once extracted from private lands or leased public lands, most oil and gas resources are expeditiously commercialized and sold on the open market where prices are shaped by the forces of supply and demand and actions of major oil-producing nations and their various regional and global associations. Royalty payments and land leasing costs aside, the resources themselves once extracted are marketable private goods. Yet, from an energy security standpoint, broadly defined, government has found a clear role for public policy; governments everywhere have chosen to treat energy, either in-use or as a result of use, as a marketable public good with significant implications for other public goods and governmental interests.

The commodification of heretofore publicly-owned and -managed resources may or may not produce positive externalities that outweigh any associated negative externalities. In other words, does shifting the supply curve to the right promote increased energy security? If so, does the increased energy security (at least in the short run) outweigh the costs of increased carbon emissions emergent from the use of these energy resources? Mining advocates frequently answer positively to the first question, and largely dismiss the second question. Meanwhile, environmental groups offer a tentative yes to the first question, emphasizing the short run, but advocate renewables as a long-term solution in response to the second and—from their perspective—the most critical question to be addressed both now and in the future.

To put things in economic terms, as shown in Figure 3.1, consumer surplus (represented as P1acP2) is created by the increased supply of petroleum resulting from oil and gas leases permitted on public lands. Producer surplus (represented as P1aeP2) is greater than the marginal private costs (MPC) (the area defined by ecd) associated with the increased quantity supplied. From the consumer and producer standpoint, the increased production of the marketable private good maximizes one important dimension of energy security—namely, affordability (a lower price for energy). Conversely, increased supply is associated with higher marginal social costs (MSC) (the area defined by abc). The increased marginal social costs might be experienced as increased air pollution, greater urban sprawl in metropolitan areas, and the related social equity and public health problems associated with those developments.

The direct negative impacts of energy development on public lands in the form of the degradation of soils, loss of native vegetation and indigenous species can also occur just as well on private lands. Contamination or depletion of aquifers as a result of oil and gas well drilling and long distance transport by pipeline or rail is a serious concern and a potential negative externality that brings with it large social costs no matter if the drilling and/or transport occurs on public or private lands. Despite the millennia-old legal precedent, the otherwise socially

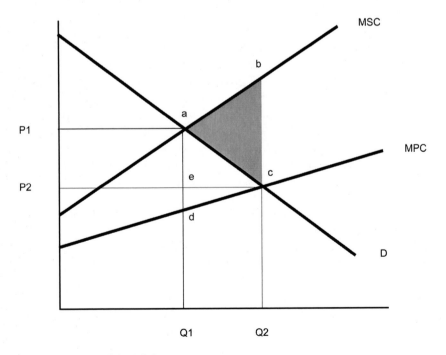

FIGURE 3.1 Marginal Social Costs

useful "rule of capture" likely invites the systematic overdevelopment of natural resources rather than promoting sustainable use. Over the long term, the depletion of scarce mineral resources will produce additional marginal social costs, despite technological market-based solutions designed to overcome petroleum and other essential mineral scarcities.

As we have seen in our discussion of a comprehensive understanding of energy security, this public policy goal entails much more than the maintenance of a ready supply for individual consumers. Historically, a readily accessible energy supply is critical to the survivability of nation-states and their sub-unit governments alike. It is in the best interests of states, regions, and nations to monitor and maintain market access to needed energy sources so that future generations are not placed at a serious disadvantage by their predecessors. A key specific motivator behind the enactment of the *Mineral Leasing Act* was the preservation of energy resources for use by the US Navy in times of national need. The US Navy had moved away from a fleet driven primarily by wind energy and had largely adopted steam-powered ships as its principal means of action. The naval vessels required a ready supply of coal and oil to heat the boilers to provide propulsion. A similar shift occurred in the British Royal Navy at about the same time. In both instances, coal and petroleum came to be viewed as national strategic resources. As a direct consequence, domestic and foreign supply sources alike were carefully protected by nations aspiring to great power status in the international arena.

The private market may not be the best method of achieving the aforementioned energy security goals. For example, the private market would not actively consider MSC without some form of constraint placed upon energy suppliers. The MSC curve is a function of governmental policy, interest group demands, experts operating in issue networks, and the prevailing values of society. Energy security lies within the MSC, and varies across energy type and source. Markets would, for the most part, view MSC of energy in relation to supply chains from point source energy development to consumers, as well as the stability of contract enforcements and the stable allocation of property rights.

Economic theory postulates that as the quantity of a resource declines, prices rise along a predictable demand curve. As prices for a good rise, lower cost substitute goods become increasingly attractive to consumers. From the consumer perspective, fossil energy depletion and use is less likely to be viewed from the standpoint of marginal social costs. Rather, it is most generally seen as an issue of supply, demand, pricing, availability of substitute goods, and associated opportunity costs.

Government, however, must take a much different view of energy supply and availability. Energy supply and costs affect national defense and infrastructure development capacity, as well as the overall competitiveness of markets and the economic capacity of business enterprises and citizens/voters. Energy accessibility and ease of use is only one dimension (albeit an important one) of national, state, or local energy goals and policies. As we have argued in Chapter 1, other critical dimensions include perceived and real levels of risk (and safety) and the long-term sustainability of supply and price stability. While risk and safety issues are likely to capture the attention of individuals and firms, the marginal social costs of energy are more likely to be dealt with through the myriad mechanisms of governmental public policy making—regulatory, distributive, and redistributive—than through individual consumer preference dynamics.[8]

Market failures provide the impetus for significant economic and social policy interventions. The protection of free and competitive markets is critical in a capitalist society, although scholars and politicians differ in how to best manage free markets and maintain a just society with opportunity to participate broadly being achieved through appropriate public policies such as free universal education through high school, favorable tax treatment for non-profit and charitable organizations helping to support an effective social safety net, etc. Scholars and elected officials opposed to large scale government intervention into energy markets, for instance, point to some glaring policy distortions properly characterized as successful rent seeking by private parties that have allowed for the creation of energy monopolies or near monopolies often characterized by highly concentrated markets, artificially high prices, and virtually insurmountable barriers to market entry.[9]

For decades, local power companies had a natural monopoly over electricity production, transmission, and distribution.[10] Petroleum, coal, and hydropower were major sources of energy used to power the generators of these power companies. In the petroleum industry, seven major oil companies—known broadly as the "Seven Sisters"—controlled virtually all of the oil and gas exploration and supply on the globe.[11] While not strictly speaking a monopoly, the oil and gas market was highly concentrated and clearly ripe for market manipulation by large petroleum companies through either collusion or the forming of cartels.

In more recent years the petroleum market has become considerably less concentrated, with multiple actors now participating in energy exploration, development, extraction, transportation and distribution. In recent years these economic forces have developed and brought to market new sources of fossil energy in the form of shale oil and natural gas on a vast scale.[12] From a free market perspective, the supply and demand forces of the free market have demonstrated repeatedly the ability to meet unmet demand through a combination of entrepreneurial vigor and market innovation. Nevertheless, the marginal social costs associated with energy security risk have worsened considerably in part due to the complex energy supply chain that has come into being to provide supplies to an ever-expanding global market for energy. For example, Kim details the costs associated with maritime oil spills in his analysis of the *Oil Pollution Act of 1990*, arguing that liability limits provided for in the law provide a disincentive for the oil industry to minimize the risk of oil spills.[13] Public policy and policy analysis have a primary role in reducing the marginal social costs of supplying fossil energy in an increasing energy-dependent global economy, despite the marginal social costs of its current usage.

While government may not directly control the level of competition present in the marketplace, it can clearly take effective action to break up concentrated markets and make the penetration of such concentrated markets by would-be competitors more feasible. Additionally, government can play a key role in providing information or requiring the provision of information to energy consumers so that they can make more fully informed choices rather than sticking solely to goods and service providers that they have used for years; these familiar and trusted sources of past supply may no longer be the consumers' optimum alternative. For example, many electrical power consumers are now informed of third party power provider options that supply only green energy.[14]

Market intervention in the form of public policy enactment is also justified in circumstances where diversification of risk is unlikely to arise in the marketplace. Weimer and Vining use the example of global warming as a risk with global implications where sufficient diversification is impossible without international-scale governmental action. Promoting zero- or low-carbon emitting energy sources is justifiable from an energy security policy because such policies are intended to reduce the detrimental impact of projected global warming in many areas of the world, including the United States.

In addition, Weimer and Vining point to the too often myopic nature of markets and the consumers' systematic misperception of risk as further justifications for public policy intervention into the operation of the marketplace. Markets are often driven by the short-term goals of clearing the market (selling goods and services) and making an immediate profit. This is a short-term focus; long-term standing is not as clear to the managers of firms whose own connection to their firms is typically contingent upon compensation based on short-term results. Similarly, the availability of some basic materials may change over time, along with the prices associated with them. Likewise, the tastes of consumers will also likely change over time, making long-term planning more difficult than if consumer preferences could be treated as a constant.

In terms of energy, a petroleum company may understand entirely that petroleum supplies will be constrained a century from now, but will likely focus much of its attention on a shorter-term goal of exploration for new oil and gas fields to meet current demand for oil industry products. Absent public policy incentives or environmental regulations, a firm that chose to divert significant resources to developing biofuels, when such fuels were only nominally in demand, likely would have experienced devastating economic loss. Without substantial public policy incentives to develop the technologies required to produce biofuels, in the form of production credits or tax incentives, the sustainability-promotion goal tied to energy security would have little likelihood of accomplishment. The costs of environmental regulation and national energy portfolio diversification could be passed on to the consumer through marginally higher prices.

Government policy might help consumers make better and more informed choices about the true costs and benefits associated with the use of fossil energy. In other words, government policy may result in more informed individual choices that result in lower demand for fossil energy, resulting in increased demand for more efficient fossil energy vehicles or clean energy alternatives.[15] As mentioned earlier, consumers generally make choices based on costs and benefits readily seen and understood. It follows that informing consumers about the true costs of fossil energy use might shape their preferences in perhaps marginal but nevertheless societally significant ways.

Micro-economic theory applied to a discussion of energy security and justifications may not be sufficient to achieving an appropriate understanding of energy security policy. As Weimer and Vining point out, utility calculations and tradeoffs can be viewed from the individual perspective, but they can also be viewed from a much broader institutional perspective. In other words, utility maximization in policies promoting energy security should also consider decisions that maximize the values of overarching social, political, and economic institutions and support the rules and processes governing those institutions along with individual-level calculations of utility.

The important dimensions of social justice and social equity require us to consider public policy from a perspective beyond that of individual and collective utility maximization. Are there policy goals that invite us to go beyond min-max decision-making? Do we at times wish to maximize benefits regardless of costs? Energy security in a society that is risk-averse about energy supply could easily justify a role for public policy in maximizing energy availability and reliability given the societal injustices and inequities that would result either from prolonged energy shortfalls or from a lack of attention to the environmental and public health consequences of unbridled fossil fuel development and use.

John Rawls approaches social justice from a unique perspective on economic equality.[16] In Rawls' conceptualization, justice ensures that the least benefitted member in society receives some benefit from societal economic gain or, at a minimum, suffers no loss. In terms of energy security policy, Rawlsian justice might be measured on several levels to include global, regional, national, and local dimensions of distribution of benefit, along with consideration for inter-generational equity. In viewing the domestic and the international energy security environment, it becomes apparent that the goal of Rawlsian justice becomes deeply entwined in multiple and at times contradictory policy dimensions.

Wiemer and Vining identify a number of "social indicators" involved in public policy analysis such as GDP/GNP, unemployment, balance of payments, and inflation; these are typical measures seen as important in understanding the degree of inequality present and assessing the general social and economic health of a political jurisdiction. From a public policy perspective, energy security studies and governmental policies relating to it often draw on the aforementioned measures to identify energy security risk and determine its possible impact on sub-sets of the nation and/or the nation as a whole. For example, concerns about the balance of payments—in this case, a measure of the economic activity between the United States and petroleum exporting nations of the world—is frequently mentioned as a reason to promote domestic fossil and non-fossil energy sources and must be considered in any discussion of energy security.[17]

The often hidden costs of energy security policy must also be considered in policy analyses. A policy choice to reduce energy security risk might result in negative impacts that are not readily apparent or typically equated to the policy implemented. For example, policies encouraging the development of domestic shale oil and the natural gas industry might result in negative impacts for local business communities and governments most directly affected by the "boom and bust" dynamics of energy development initiative in the past.[18]

Identifying a policy need based on public goods impacts or social justice considerations is only the beginning of a very difficult process of more specific problem definition and solution identification and ultimate policy choice. Policy prioritization can easily be distorted in the ranking process along the lines of Arrow's *Impossibility Theorem*.[19] In representative government and in a federal system such as the United States, there are multiple veto points within the legislative process as well as the pressures posed by economic and public-interest groups, bureaucrats, and technologists. Interests may operate individually or in unison through narrowly defined issue networks. In the case of energy security, interests and issue networks may be composed of international and sub-national governments, cartels, or NGO actors. Networked and independent interests alike are able to gain timely access to legislative representative and bureaucratic decision makers with the intent of shaping the public policy agenda, influencing issue framing, generating solution choices, and affecting implementation processes to assure continued access and influence long after specific legislation is enacted.

In US energy security policy—broadly speaking—the actions of the multiple actors typically involved from international and subnational levels of government may combine to distort public preferences in such a way as to reduce rather than maximize public policy goals. Emergent policies may serve to maximize the benefits of particular interests or policy network actors' agendas while the associated costs are broadly distributed across society and hence are more difficult to appreciate and to muster support for effective policy advocacy. For instance, in the aggregate petroleum remains a necessary good in the industrial and post-industrial world, demand being largely inelastic. Policies that increase the cost of petroleum are relatively easily passed on to the consumer in the form of marginal increases in liquid or gaseous fuels, often to the detriment of other sectors of the economy.

The petroleum dilemma illustrates well a constructive role for policy analysis, but highlights as well the potential limitations to be recognized. The political roadblocks to effective policy analysis are numerous and require consensus-building on multiple stages—on problem definition, on policy formulation, and on policy implementation. It is not clear that the results of consensus building will result in a rational and fair outcome—one that maximizes benefits in relation to costs (to include the maximization of energy security goals) while simultaneously producing Pareto outcomes. In light of growing substitute energy products, such as shale gas, analysis of historic and current energy security policy—and reflections on future need—must regularly consider marginal social costs in relation to marginal production costs, regulatory costs, demand and supply elasticities, and market prices. From a free market perspective, the MSC associated with energy security are politically defined, but the costs of reducing MSC are paid for in real dollars by the very consumer that public policy is designed to protect. At least in the short term, the net benefit to the consumer of timely energy security public policy intervention is, therefore, difficult to detect.

Policy analyses of energy security must do more than define typical best public solutions. Extrapolating from Aaron Wildavsky's *Speaking Truth to Power*, the argument could be made that policy analysis should also help elected officials and citizens better understand ways in which energy resources might best be protected.[20] This would require the policy analysis entailed to articulate the risks, identify the costs, and document the benefits associated with various approaches to creating and maintaining sustainable communities which best meet our needs and protect public goods to the fullest extent possible with full recognition of inevitable tradeoffs. Greater rationality in decision-making as a result of good policy analysis would help us to articulate more clearly our normative values upon which final choices are made, as well as properly locate them in the policy process. As noted in Chapter 2, societal values do indeed shape definitions of risk, affect problem definition, and influence policy choices; and, clearly articulated values lead to policy choices that go well beyond simply meeting the minimum requirements of protecting public goods and promoting Pareto outcomes in policy and society.

Policy analysis should also help elected and appointed officials and the broad array of citizen and corporate policy stakeholders to produce outcomes that do not result in the overproduction of public goods at the cost of long-term sustainability. Good policy analysis will also aid in making good decisions on the part of government about how best to use policy to shape prevailing market forces—public policies that dissuade harmful market concentration and foster healthy competitive market development, providing timely openings for new energy providers using renewable sources of originating energy.

A hopeful scenario emerging from successful policy analysis must, however, balance present and future needs with the short- and long-term impacts of sunk costs on energy security policy. The sum of all sunk costs related to energy development, transportation, and effective use are perhaps the single largest roadblock to rapid changes in the historic direction of energy security. The scale of sunk costs in the energy area is great indeed, shaped by a combination of private sector choices and public policy frameworks within which are articulated a variety of implementation rules and processes advanced by policy incentives and regulatory structures.

Energy Security, Property Rights and Marginal Social Costs

Chapter 1 outlined a very broad definition of energy security. Given this understanding of the term, we find that US policies related to energy security cover a rather large scope of public laws, court cases, and policy implementing public agencies. Energy policy developments in the 20th and 21st centuries generally sought to reduce the marginal social cost of energy production and use, but have in the process a clear tendency to reduce property rights for individuals and corporations in policy areas that have lost political ground to "green politics" interests and environmental protection advocates.

A property rights-centered dialogue was more common in the 19th century, largely because government institutions—either through new statutory authority or through established common law precedent—were only beginning to articulate the concept we refer to as marginal social benefit (and marginal social cost), key components in understanding the core dimensions of contemporary energy security policy. While negative externalities can be identified and articulated in nearly all private and public sector choices, the markets giving rise to them were by no means as expansive as they were later to become in the latter half of the 19th century. With the emergence of the industrial capitalist state in the 1880s, paralleled by the dramatic growth in the size of cities, awareness of the emerging negative externalities of industrialization and urbanization became more widespread. The property rights dialogue alone was broadly seen as by no means capable of managing the dilemmas surrounding the imposition by the industrial state of social costs and the resultant infringement of individual rights. In fact, it wasn't entirely clear how and if government could (or should) identify and seek to minimize social costs and promote individual rights to life, liberty, and the pursuit of happiness.

In terms of energy security and marginal social benefit, perhaps the earliest example in US government operations comes from the case *New Jersey Steam Navigation v. Bank of Boston*.[21] The case involved a steam-powered shipping company and illustrates well the role of energy source in an understanding of energy security in terms of social costs. Chartered in 1839, the New Jersey Steam Navigation Corporation had used wood charcoal as a fuel source to heat its boilers to produce steam to then power the locomotive devices of its ships. Wood charcoal was a manufactured product that involved the deforestation of many acres of land along the Eastern seaboard and into part of the interior. Charcoal making was a labor-intensive process conducted by individuals known as colliers—charcoal makers. Known as the process of pyrolysis, the chemical decomposition production procedure involves charring wood and accomplishing the removal of resinous matter and water all in a low-oxygen, high temperature environment. In the early 19th century, the process could take several days to complete and was a seasonal industry. As land was cleared for development and agriculture, the distance between collier operations and markets grew, resulting in increased costs of transportation.

In the late 1830s the retail cost of charcoal was approximately $6.90 per ton. While coal ranged in price from roughly 0.66 to 1.5 times the price of charcoal, the heat value of coal was more than twice that of charcoal while taking up significantly less storage space on board a ship. Therefore, from an economic standpoint, charcoal became an inferior good in comparison to coal. Shipping companies such as the New Jersey Steam Navigation Corporation began to switch from charcoal to coal in order to maintain or grow their profit margins. In the process of making the switch in fuel source, however, the company did not realize that it would require major upgrades to their furnaces and boiler rooms to accommodate a substantially higher temperature fuel. A shipboard fire harmed numerous individuals on board (with some losing their lives) and destroyed all the property being shipped.

Under its admiralty jurisdiction, the US Supreme Court found itself in a position to establish a policy on energy security and safety based upon American law. The Court found the shipping company liable for the destruction of property and for loss of life. Furthermore, the Court drew on the limited maritime safety statutes of the time to frame its decision. In other words, the US Supreme Court used not only its interpretation of public good and social benefits attached to economic endeavors, but it relied also on public law produced by the elected branches of the federal government. Finally, the court stayed within the property rights narrative, but by discussing the issue of potential person-related criminal actions (manslaughter resulting from corporate negligence), the Court either wittingly or unwittingly opened the door to a broader discussion of social justice-related issues tied to energy security. While the latter would not become fully evident for some time, it is interesting to see the US Supreme Court, as a governmental institution, begin to grapple with the issue of marginal social costs in relation to marginal production costs and thereby recognize the need for public policy to articulate the manner in which social costs will be identified and managed.

The *New Jersey Steam Navigation* case also illustrates the origin of an often-misarticulated distinction between energy security and energy safety. As illustrated by the case, market forces led to the large-scale substitution of wood and charcoal with anthracite coal as a fuel source. The latter was a superior product offered at a lower cost. While wood was plentiful, charcoal manufacturing was time-intensive and the costs of transportation to market were rising. Price signals likely led to the wholesale switch from wood and charcoal to coal in a largely unfettered free market. The court's intervention into the market in this case was a result of energy safety issues. The US Supreme Court established a precedent in US law carried forward to this day in linking liability to product used in providing a good or service.

Transportation fuel usage changed in the 19th century, as witnessed in the aforementioned case, but the early 20th century brought another new dramatic shift—namely, the widespread development of a new form of transportation in the form of the automobile. In 1886, Karl Benz patented the gas-powered automobile; since then, the energy world has never been the same. A few years earlier another technological breakthrough had changed energy usage in lighting—the invention of the light bulb by Thomas A. Edison, and the opening of the first electricity generation stations in Europe and the US. As a result, there was now and remains so to this day, a huge demand for energy sources to fuel these two particular inventions: a demand for liquid fossil fuels (diesel and gasoline), coal, and natural gas.

Eastern states, such as Ohio, Pennsylvania and West Virginia played a prominent role in providing petroleum, natural gas, and coal to feed new forms of transportation and for a growing industrial base in the United States. In many cases, mineral rights to subsurface resource development were sold or leased by private landowners to companies with an interest in developing the energy resources. In the American West, much of the land was owned by the federal government; consequently, developing mineral rights claims functioned a bit differently because the mining interest was dealing with the Sovereign, which had both property rights interests to consider and an evolving notion of the public good or public interest to factor into its actions.

Much of the American West had been gained through conquest in the victory over Mexico in the War of 1848. Shortly after the war's conclusion, gold was discovered in California, followed by a rush to stake mining claims and to unearth the precious metal. The process by which claims were staked was haphazard and was made more challenging by the fact that the federal government, having not anticipated the property rights issues, now faced great difficulty in trying to dislodge an anxious and at times violent group of miners occupying public lands. There were attempts to regulate mining on the newly acquired lands, but the attempts were largely half-hearted efforts within the territories; politics at the national level was increasingly focused on a bigger issue—namely, an impending and subsequent Civil War. The Northern States could hardly afford to rein in mining claims on public lands in states that would play a key role in legislative compromises between the North and the South, lands crisscrossed by a transcontinental railroad built by individuals with a keen interest in mineral development on public lands, transporting people, precious metals, fossil energy, and other natural resources across the continental landscape to feed resource-hungry cities and industries in the East.

The Federal government's landmark attempt to regulate mining on public lands came with the passage of the *General Mining Act of 1872*. The act firmly established the process of staking mineral claims on public lands, and remains largely intact to this day except for some key restrictions on fossil energy development. For several decades, fossil energy mining claims on public lands were limited as coal demand was predominantly related to heating homes and offices or running steam engines for transportation or in industrial applications. All of this changed due to a sharp increase in demand for petroleum, natural gas, and coal to fuel the new transportation and the age of electricity. Petroleum, natural gas, and coal claims on public lands increased dramatically, causing deep concerns about the depletion of strategic national resources, a serious energy security concern of the era.

In response to this security concern, President Taft's Temporary Petroleum Withdrawal No. 5 Executive Order cordoned off over three million acres as a potential energy supply source for the US Navy. President Taft's move to restrict private mining claims was not entirely unprecedented, however. In earlier decades, other public lands had been designated for various public purposes in the national interest. Native American Reservations had been created with this logic in mind. National parks had been designated by President Taft's predecessor, Theodore Roosevelt. And in 1897 the *Sundry Civil Appropriations Act (Organic Act of 1897)* laid the foundation of public forest management on federal lands. The *Antiquities Act of 1906* championed a growing effort to preserve irreplaceable natural beauty on public lands. A growing concern with the timely preservation of noteworthy public lands (e.g., the Yellowstone area, Yosemite, the Grand Canyon) and an evolving conceptualization of the public good had important implications for property rights allocation in mineral energy development. There was a growing awareness among Americans across the country of the marginal social costs of non-renewable resource extraction.

In 1910, Congress passed and the President signed into law the *General Withdrawal Act*. The law formalized the use of withdrawal in conserving public lands, "authoriz[ing] withdrawal of lands of historic and scientific value; limit[ing] size to the smallest area compatible with management for these values."[22] The *Stockraising Homestead Act of 1916* allowed for the issue of "surface land patents that reserved all subsurface minerals to the United States."[23] The early 20th century movement towards redefining and narrowly constructing property rights claims on public lands culminated in the *Mineral Leasing Act of 1920*.

The *Mineral Leasing Act* separated surface property rights from subsurface property rights. Mineral leases could also be limited in number, location, size, and duration. The mineral resource being mined was owned by the government, and the government could manage that resource, continuously supervise its withdrawal, and place restrictions on its processing. Government could also (and did) collect royalties, a percentage of gross revenue generated, assess rents, and impose other fees associated with the accessing of a mineral lease on public lands.

More importantly, the *Mineral Leasing Act* gave the Secretary of the Interior the power to place limits on leases. The Secretary was given broader authority to determine the legitimacy of a claim. The Interior Department could and did police the terms of lease agreements to ensure that claims once permitted were acted upon during the prospecting period (up to 2-year lease). Land leased to an individual or corporation intent on prospecting, developing, withdrawing and selling the minerals withdrawn (e.g., coal, oil, phosphates, natural gas) could not exceed 2,560 acres. After "valuable deposits of oil and gas have been discovered" the entity holding the permit was "entitled to a lease for one-fourth of the land embraced in the prospecting permit."[24] The royalty payment for the oil and gas production lease was set at 5 percent of gross revenue. The remaining land in the prospecting lease could also be developed by the prospecting lease, but was subject to a competitive bidding process with a minimum royalty rate of 12.5 percent of gross revenue. Coal carried a royalty of 5 cents per ton to not less than 12.5 percent in the *Federal Coal Leasing Act of 1975*.[25] The law states that production leases were valid for up to 20 years, so long as production was occurring at a commercially viable rate. There are explicit reporting requirements by the lessee to the Bureau of Land Management (BLM). Furthermore, BLM must produce reports to Congress on the uses being made of federal lands and the revenue generated from the development of the land. Ultimately, the resources developed from the lands have always been intended to benefit the United States as the trustee for the People; in fact, ownership of the lease cannot be held by a foreign national—neither direct or stock ownership, but this restriction does not prevent the mineral resource extracted from being sold on the global market post-resource recovery.

The *Mineral Leasing Act* has been amended substantially several times since 1920. The provisions of the law have evolved in direct response to changing technology and changing patterns of resource usage. In 1958, the law was amended to cover mineral prospecting and the development of leases for offshore drilling. The *Mineral Leasing Act* was further amended as a result of the *Geothermal Steam Act of 1970*. Technological breakthroughs in geothermal energy production led to geothermal developments on both public and private lands. The *Trans-Alaska Pipeline Authorization Act of 1973* involved amendments to Section 28 of the *Mineral Leasing Act*. The 1973 Act established rights of way across federal lands "for the transportation of oil, natural gas, synthetic liquid or gaseous fuels, or any refined product produced"[26] to the lower 48 states. In a bow to a changing policy environment and to the growing complexity of mineral lease development, the 1973 Act clearly recognized the environmental protection role of government enshrined in *National Environmental Policy Act (NEPA) of 1969*, and emphasized the need for systematic interagency and intergovernmental collaboration.

In the roughly 70-year period between the *New Jersey Steam* case and the passage of the *Mineral Leasing Act*, the direction of public policy had shifted tremendously in the United States. Energy security had moved from issues of energy safety to issues of energy security in terms of resource supply and ownership of energy supply. The public goods aspects of energy in the mid-19th century had focused on issues of liability in energy use, particularly energy use and its potential impact on private property and human life.

The public goods dimension, however, required a full re-articulation in the period between the conclusion of the Mexican–American War (1848), and more particularly in the period following the placement of restrictions on public land uses. The initial issues arising came in the form of challenges by pre-*Mineral Leasing Act* claimants to the federal government's ability to change the rules governing the exploitation of public lands for private gain. In the case *Wilbur v. Krushnic* (1929), the US Supreme Court found that enforcement of *Mineral Leasing Act* lease standards did not apply to individuals holding mining claims issued prior to the *Mineral Leasing Act*. In other words, the Interior Department could not, in effect, use the new law to invalidate pre-*Mineral Leasing Act* claims using *Mineral Leasing Act* standards (e.g., the $100 per year expense for mining-related activities on the claim). The basic principle within *Wilbur* was effectively reaffirmed by the US Supreme Court in *Ickes v. Virginia-Colorado Development Corporation* (1934).

In *McClellan v. Wilbur* (1931) the Court reaffirmed that the Secretary of the Interior could limit application approvals for mineral development based on reasons other than following the leasing application process. Oil and gas applications had been denied by Interior Secretary Wilbur for reasons of the general welfare. In the early 1930s, oil and gas demand was depressed due to the dire economic conditions prevailing in the United States at the time. The oil and gas revenues generated would have been small, and the glutted markets would have been further weakened. In consideration of these public interest matters the permits requested were denied. The US Supreme Court specifically addressed the issue of resource demand conditions in their opinion sustaining the actions of the federal government in this case of permit denial.

McClellan's focus on general welfare is a theme echoing from state court decisions of the time as well. In *McNeill v. Kingsbury* (1923), an oilman challenged California state law governing petroleum claims, as covered in the state's *Oil Lease Act*. The oilman had sought a permit to explore for oil on property owned by a state mental hospital. The state surveyor had denied the permit on the grounds that the property was already serving a state interest and that oil exploration on the property would disrupt the state interest. The California Supreme Court found that the land had been reserved for a "special purpose" and that oil exploration and development would be disruptive to that purpose as maintained by the State of California.

What emerges here is an early attempt by American courts to build a new legal conceptualization of social costs related to energy resource development. These costs related to something other than energy safety and associated product liability issues. Clearly, it would be possible to develop petroleum and gas resources despite low market prices. From a free market perspective, so long as marginal benefits equal marginal costs, production of the good or service will proceed in the short term. Yet, the American courts of the day held that market principles do not provide a final source of guidance in the case of mineral leases—there is rather a higher good to be served, which is the public welfare. The public welfare includes getting a fair return in the form of royalties paid, reflecting the worth of a non-renewable resource on land owned in trust by the government for the people.

It was not until the 1930s that the Court took on a broader range of cases involving energy security and safety issues. This was part of a larger shift in which the federal government came to assume greater responsibility for the national economy and the economic welfare of individuals. The landmark case of *Lochner v. New York* in 1905 enshrined the view that government could not regulate working conditions (in this case, the maximum number of hours a day laborers could be required to work), as it was a violation of the implicit "right to contract" in the Fourteenth Amendment. Over the next three decades, the Court would issue a number of opinions that invalidated Progressive Era federal and state laws that governed the workplace and by extension, the functioning of the economy. However, it eventually began to shift. The precedential case of *West Coast Hotel v. Parrish* (1937), which effectively repudiated the view taken in *Lochner v. New York*, took the Court down the road of permitting the government to regulate aspects of the economy. Plaintiffs in the *West Coast Hotel* case argued that the state law imposing a minimum wage violated the Fourteenth Amendment—requiring the hotel to forfeit property without due process of law and interfering with private contract arrangements between the hotel and its employees. From a security and social justice dimension, the key points of law in the case deal with the issue of economic security for a protected group in society; in this case, the protected group was women in the hotel workforce. The US Supreme Court found that there was no discrimination against West Coast Hotel because the law applied to all businesses and would, therefore, not limit the ability of the hotel to compete for customers.

The basic principles emergent from *West Coast* had broad implications for energy security and safety policy. In a case occurring five years prior to *West Coast*, the Court took a stab at fossil energy environmental regulation. In *Champlin Refining Corp v. Corporation Commission of Oklahoma, et al.* (1932), the Court struck down broadly written state-level environmental regulations governing the petroleum industry. The US Supreme Court opined that environmental regulation might be required in some circumstances, but it must be narrowly constructed and built on solid scientific evidence. While the court's decision set aside state-level environmental regulations, the very fact that the Court took these cases and set forth the criteria for acceptable regulations effectively places state economic and safety policy within the scrutiny of the national government. Perhaps the most significant thing to consider is the combined effect of these cases— namely, the importance of narrowly tailored public policy. In the *West Coast Hotel* case, narrowly tailored economic policy focuses on equitable *group* rights without distinction to economic or social class—namely, the rights of women in the workplace. In *Champlin*, occurring five years earlier, the Court places greater emphasis on narrow construction of policy regulation and a narrow construction in the articulation of marginal social costs. One lesson to take away from this review of precedential cases is that the articulation of individual rights moves from narrowly to broadly constructed classifications. Articulation of marginal social costs may expand independent of the marginal production costs, marginal increase in supply of a good that has incorporated within its production, transportation, exchange or use some form of intentional or unintentional marginal social cost.

What is also to be gleaned from the Court's choice of energy-related cases is a prominent role accorded for cases involving negative externalities. Scientific evidence and changing societal values produced the watershed policy shifts of the late-1960s and 1970s. The *National Environmental Policy Act*, the *Environmental Protection Act*, the *Clean Water Act* and the *Clean Air Act* (and 1990 Amendments) greatly changed the policy environment for energy-related activity. It reshaped and more carefully articulated the rights-based issues surrounding private market activity; in this case, the energy industry writ large. While key legislation created the major regulatory institutions and set broad national goals, the growth of advocacy coalitions bringing together scientists, interest groups, and regulators—operating in a revolving door environment involving service in government at some point—clearly helped to articulate the public goods aspects of energy policy. Key administrative rules at the federal and state level were crafted by members of this broad, loosely connected network of deeply knowledgeable individuals working in government, nonprofit advocacy groups, academia, and in private consultancies.

Much as the case with petroleum and coal resources on public lands, electrification has witnessed an evolving articulation of property rights issues, identification of marginal social costs (and benefits) related to its production, transmission, and use. Electric power provides a good example of how conceptualization of energy security and safety can be reshaped in a major way over time.

In the early decades of its development and commercial and household use, electricity was not viewed as a strategic good and was, therefore, not actively regulated at the national policy level. Electricity was seen as an exclusively local or regional concern. Electricity markets were, for the most part, regulated by state and local public utility commissions in an attempt to reduce market inefficiencies created by natural monopolies. Nevertheless, electricity supply was limited, demand was growing, and prices made electricity inaccessible to many individuals; this was particularly the case for individuals living in rural areas.

National level politics and policy began to develop in the 1930s. At this point, electricity began to see its role in the national economy shift in a major way. With World War II looming in America's future, electricity came to be seen as a strategic good for war industries. Electrification policies also took on a distributive policy function, moving the good from a marketable private good to a marketable public good—energy security was brought into the larger social safety net being cast across society in the form of *New Deal Era* policy. The majoritarian politics-themed New Deal policies blended well with a continued majoritarian theme promoted decades later by environmentally-conscious advocates for renewable energy.

Renewable energy advocates, however, have articulated a very different idea about "public goods." What are the public goods in question? Among the most prominent public goods identified are those of environmental protection, social justice, regular supply of energy resources, and sustainability broadly understood. Environmental protection is perhaps the most prominent public goods concern, effectively moving the public goods debate to interspecies, international, global concerns. At these levels, energy security issues moved beyond that which local, state, or national public policy could effectively regulate and manage. Individuals, groups, and nations advocating for a wider global environmental politics and policy framework for the advancement of energy and environmental security required a global-sized policy agenda to advance their cause. This time, however, green energy policy advocates saw an opportunity to bring together two different issues in a way that made it possible to articulate the distinctive and shared aspects of both, demonstrating the need to promote both issues simultaneously via public policy. Those now-linked issues are *energy security* and *energy safety*.

This global-sized agenda, which has sought to address sustainability and the negative externalities in an international context, has been ambitious. In 1990, the *Report of the Intergovernmental Panel on Climate Change* (IPCC) linked the intense use of fossil energy and related human activity directly to climate change. The landmark report warned of dire consequences for humans and other species. Rising temperatures would lead to increased aridity in large parts of the planet. Polar ice caps would melt down and lead to rising sea levels, and the flooding of low lands around the world—densely populated areas as well as small island nations—would affect millions of people throughout the world. Weather patterns were predicted to become more extreme, causing widespread property and environmental damage from uncommonly severe storms. The report was quickly followed by an international conference on climate change held in Brazil.

The 1992 United Nations Conference in Rio de Janeiro brought the shocking findings of the IPCC report into an international political arena. Developed and developing nations alike were called upon to begin to engage in a broadly-based dialogue about the future of the planet. A significant part of the discussion revolved around the climate crisis challenge—the widespread use of fossil energy. Climate altering emissions—carbon dioxide, sulfur dioxide, and reactive nitrogen—are all related to the use of carbon-based fuels. Energy security was no longer simply a matter of readily available and affordable supply. It was also a matter of managing the negative environmental externalities of fossil fuel energy use.

The Kyoto Protocol signed in 1997 by 83 nations established carbon emission caps to reduce the climate-altering impact of fossil fuel use. For the first time in history there was significant global agreement that energy security was not just about the supply of energy, but also had to do with the multiple negative externalities produced by the widespread and growing use of particular forms of energy—namely, carbon-based energy sources.

Social justice has been prominently linked to the issue of climate change as that issue has developed in the research and environmental policy studies literature. From a more universal perspective, climate change wrought by human activity carries with it a penalty for all living things on the planet; humans of all social, economic, cultural, racial, and ethnic backgrounds figure prominently in the issue as framed. Yet, the human activity that led to potentially deplorable conditions is not a function of equally shared guilt or blame, but rather it is predominantly a function of highly developed post-industrial nations, such as the United States and the EU nations, and increasingly a problem of large rapidly developing industrial nations such as China and India.

The regular supply of certain forms of energy resources is also seen through the lens of public goods, albeit it is generally recognized as a marketable public good. Electricity is particularly viewed from a public goods perspective, linked to many other related concerns about human rights and democratic governance. A study of residential electrification of sub-Saharan Africa linked electrification with the presence of institutions of democratic governance. A basket of public goods—basic human rights and the right to self-governance—is therefore linked to the security of energy provision. In other words, electricity (produced from no specific source) is statistically linked to energy security, but that security is at the hub of a wheel linked by implication to democratic institutions and to free market conceptions of property rights and social costs and benefits, broadly defined.

Economic sustainability is also prominently linked to energy security. It is couched within a broader set of criteria relating back to issues of social and intergenerational justice, environmental justice, public health protection and myriad other do-no-harm constraints that exist under the green politics and/or sustainability umbrella. The maintenance of Pareto efficiency as conceived by writers within this tradition is not seen to occur through classic free market transactions, but is flattened and stretched beyond the seller and buyer to include all of society in which private and exclusive transactions are limited and the market is subsumed by a broad conceptualization of governance—in other words, all for the benefit of all.

Renewable energy development poses new dilemmas for property rights. Just as there were created a myriad of property rights rules governing extraction industries operating on public lands, renewable energy developments on those same public lands are open to further consideration. As has been alluded to earlier in the chapter, the exercise of property rights governing energy extraction often seems to function independently or semi-independently of the regulations governing identified marginal social costs. Social costs that may not have been articulated at the time of a mineral lease being established or exercised may, at a later time, become the basis of policies regulating energy extraction and become the basis of a defined energy security risk. The social costs are derived either through scientific evidence, or through statutory, administrative, and/or common law precedent. More importantly, the direction of government regulation of property rights for reasons associated with marginal social costs has expanded tremendously, particularly in areas governing environmental impacts of energy resource development, manufacture and use.

While fossil energy development saw marginal social costs from resource development grow, renewable energy development tends to work in the opposite direction. For example, if an individual decides to put solar panels on their home's roof, but their next-door neighbor makes a landscape or building decision that shades those solar panels hence making them less effective, statutory, administrative and common law precedents tend to protect the rights of the renewable energy advocate to the disadvantage of the neighboring property owner. The exercise of private property rights (the neighbor's) becomes a marginal social cost to the renewable energy advocate and the greater society interested in clean renewable energy. Unlike cases involving protected wetlands, for example, such cases of disadvantaged property owners do not involve a direct impact on a protected species or habitat through the denial of their access to a defined resource of limited availability.

In terms of property right precedents governing mineral resource extraction, the "rule of capture" has meant property rights are governed by getting to the resource first. In the case of renewable energy, therefore, if an individual or corporation were to build a huge renewable energy facility on a large tract of land and erect structures that shaded nearby solar panels or altered wind patterns to the disadvantage of wind power collectors, then said individual or corporation would be wisely following the principles embodied in the "rule of capture" by obtaining a property right to a well-situated site for renewable energy development. In fact, contemporary government regulation of public lands and associated renewable energy facility leases follows this principle of long legal standing; residential renewable energy development, however, relies heavily on creating winners and losers in terms of the exercise of private property rights, creating a new set of marginal social costs to some homeowners.

Besides property rights governing land use, water rights also come into play. In energy resource development and extraction, water rights are as important if not more important than obtaining access to a land resource. Water rights, mineral claims and land leases, and land ownership and real estate ownership and land use in the arid Western United States were seemingly born together largely because of their mutual importance in the exercise of property rights. The *doctrine of prior appropriation* governs water rights in the American West. Without going into great detail here, it simply means that those who first make a claim to a water right have first priority in its use:

Under the prior appropriation system, the rights of latecomers were sub-ordinated to those of early users. Nearness to a stream was not considered a factor in the order of priority rights .... Often, downstream users have claims senior to upstream users, creating problems that involve the accuracy of records and the means of ascertaining ownership rights, as well as determining the prior rights to water that may be held by the federal and state governments as well as by Indian tribes.[27]

A property rights dilemma that emerges for energy security in the promotion of renewable energy sources is that, assuming that renewable energy from solar, wind, or biomass sources is a viable or highly desirable solution for the promotion of energy security, how does a private energy firm make claim to needed water resources when more senior water claims made by other non-government inter-ests hold senior claims to said water? Either the water right is obtained through purchase or, particularly in the case of public lands or Native American reserva-tions, federal reserved water rights are exercised and take priority over other private or sub-federal government water rights claimants.[28]

Marginal social costs being a determining factor in regulatory decisions gov-erning the access, development and use of natural resources, and given that such costs are measured in terms of financial costs, a marginal social cost is an eco-nomic cost to a person or corporate entity; and policies seeking to overcome such marginal social costs, at times, produce a marginal social benefit in the form of an economic benefit to individuals or corporate entities. In the case of water rights and water use, the act of prioritization for renewable energy development, parti-cularly on public or Native American lands, has implications for a variety of types of property rights, and litigation over these water rights conflicts could go on for decades, as has been the case historically. Water rights reform has been shown to be incremental and evolutionary, just as public policy is often properly characterized as evolutionary in so many respects.[29] The evolution of prior appropriation doc-trines and ground water use has slowly adapted to societal change over time to meet new and changing demands given changing conditions.[30] In the case of sub-surface water this natural resource has been subject to a combination of local, state, and federal laws that seek to manage the resource, particularly if it is not rechargeable or has been shown to be very slow to recharge.

Energy Security from a Market Perspective

Perspectives on energy security tend to vary across regime, energy type, and market characteristics. Regardless of security and safety issues, energy is viewed by many major petroleum producing nations as a form of marketable public good. The protection of energy supply and disposition rates are viewed as paramount energy security issues. In the Organization of Petroleum Exporting Countries (OPEC) nations such as Saudi Arabia and the United Arab Emirates, petroleum and gas resources are the property of the kingdom or emirate, respectively, until the time they are sold to private concerns. Even after their purchase the use and resale of those resources are subject to prior agreement between Saudi ARAMCO or the Emirate Ministry of Energy. From the supplier perspective in this case, the resource provides for the basic sustenance of the entire nation's population and its governing class. In state-owned petroleum and gas enterprises, the energy resources are seen as a common pool resource to be managed in the interests of the state. Sustainability in the sense discussed previously is articulated only in cases where resource depletion is on the relatively near horizon; those with more extended horizons tend to focus greater effort on resource release rates either in order to shape favorably market price structures or to respond to global shortfalls—the latter being largely the reserve of Saudi Arabia as the only nation with sufficient oil and gas reserves to deal with the issue effectively.

The United States takes a somewhat different perspective on its petroleum resources. Resources can be privately held either as a result of private acquisition or through claims staked for resources beneath public or private lands. While resources beneath public lands may be limited in development due to environmental restrictions, the resource and its development is generally viewed from a market private good perspective. With the exception of restrictions on resource development for strategic purposes and advancement of environmental and other federal land use priorities, the United States government has largely divested its ownership of sub-surface mineral rights. In the United States, state and federal agencies regulate the process of extraction and seek to minimize negative externalities associated with the recovery process; but the amount of petroleum or gas that is extracted is not under government purview, a fact which has important implications for federal policy governing energy markets, increasing reliance on the articulation and regulation of marginal social costs as a method of increasing energy prices in the hopes of constraining demand of certain nonrenewable fossil energy sources. For example, environmental regulation costs could be a mechanism to increase the costs of shale oil and gas production, the former already facing higher production costs due to the existence of only limited refining capacity for heavy shale oil.

In the United States, the architects of both the *Rural Electrification Act* and the *Public Utilities Holding Company Act* recognized a similar basic need in modern communities and economies. Electricity supply, transmission, sale, and access were to be regulated in recognition of this necessity of modernization. In this instance, however, market regulation was used not to reduce access or increase prices; rather, it was intended to provide energy security through greater access to energy supply and to regulate (in most cases moderate) consumer prices in often monopolistic market settings. Regulation of electricity establishes this vital energy source as one that possesses at some level the characteristics of a marketable public good or common pool resource. Electricity is seen as fundamentally different from petroleum and gas—the former cannot be stored in significantly large quantities and lacks portability beyond the transmission grid, whereas the latter is by its very nature stored energy that is highly portable over surface or near sub-surface transportation modes (roads or pipelines) and thus is marketable globally.

Why is any of this prior discussion of energy as a market good important to a discussion of energy security? First, the nature of a good shapes the ability of government to regulate that good. Second, the nature and scope of a marketplace for a particular good affects the ability of government to regulate or shape both the market and the goods being bought and sold and to achieve regulatory policy goals. As discussed in the first chapter, energy security is primarily concerned with the regular supply of an energy resource, its price fluctuation, and the environmental externalities associated with it. Only in the case of electricity was the market a natural monopoly that lent itself to government regulation of the market. This turned out to be a precondition that is subject to dramatic change due to the advent of third party power contracting, grid regulatory reforms, and off-grid power production and usage on a considerable scale.

In most other nations, energy resources such as oil, natural gas and coal that are extracted from the ground are owned by the government, not private companies or individuals, and are themselves treated as marketable public goods not subject to private trading. By way of comparison, the United States, with some exceptions, allows for private ownership of energy resources in its unique system of property rights, and treats energy sources as marketable goods narrowly defined. The marginal social costs associated with the use, extraction, production, and transportation of the resource are the basis of the public goods dimension of energy.

In nations where governments own the energy resources, government has a great deal to say about the development of the resource and its extraction rate. Saudi Arabia, for example, might lower its pumping rates if the price of oil dropped substantially. National revenue demands play a role in shaping pumping decisions as well in the case of Saudi Arabia and other oil-rich nations. At first glance, this seems like a good way to control and shape an important dimension of energy security; namely, a regular supply of an energy resource sold at a price that is within an acceptable range.

In the United States, however, policy reflects the proposition that problems emerge from national ownership of the mineral resource. When a democratic government retains direct ownership, development, and sale of a marketable private or public good to be made available for private use, particularly one with tremendous power to shape markets, it is no longer acting as a neutral party in the regulation of a market. In such cases that government becomes a competitor, and often directly competes with individual or corporate actors in the marketplace— marketplaces that government itself regulates. The potential for negative impacts on free operation of energy markets can be significant under these circumstances.

The future of energy security policy is unknown given the multiple uncertainties at play; yet, there exist important clues as to its most likely general direction. Some of these clues can be found in the nature of energy transitions—that is, the substitution of one form of energy for another over time. In studying energy substitution over the last century, it is interesting to observe that substitution can occur in a relatively short period of time. A good example of this can be seen in the transition from whale oil to electricity and natural gas for domestic illumination. In the 1850s, whale oil lamps were a primary source of lighting in most homes. By the 1860s, natural gas supplies were being piped into towns and emerging cities to be used in home and office lighting as well as for cooking and heating. Early public utility commissions were tasked with regulating the new fossil energy sources in terms of availability, safety standards, and price. Planning divisions within municipal governments worked with natural gas companies eager to install supply lines under municipal streets. By the turn of the century, electricity was more commonly found in cities and towns and light bulbs were manufactured in large quantities. Over a period of three decades, whale oil demand fell, although whaling continued as demand for other whale products remained quite robust. In this case, the energy substitution initially occurred as a result of free market innovation; it wasn't until the 1930s that government energy policy effects became more pronounced in the area of electrification, reducing consumption of kerosene used for lighting in most rural communities.

In the former case, policymakers facilitated a market innovation. In the case of renewable energy, government policy by-and-large created a market and fed innovation through distributive and redistributive policies, often using its regulatory power to weaken competition from the nuclear and/or fossil energy market sectors. The combined effects of atmospheric emissions standards on the fossil energy production costs (e.g., stringent air and water quality standards applied to coal power plants requiring carbon sequestration) and the creation of carbon markets and renewable energy emissions credits have incentivized a move towards energy source transition. Whether that change would have occurred more slowly without these market-altering actions is unknown, but if history is any guide it is likely to have occurred in its own way over some period of time. Just as public utility commissions (PUCs) in the 19th century made room for natural gas in towns and cities across the United States, public policy at the national, state and local levels is actively shaping energy markets, moving the market in new directions without, however, declaring energy per se to be a public good to be meted out directly by government.

There is a great deal of value in the market-based mechanisms behind the transition from one energy source to another. Through a series of individual market transactions, the relative quality and security of particular energy sources can be gauged. Market actors involved in energy transactions use prices to send signals about the relative demand for a particular good. The specific use of the energy good may be either known or not known, but this information may be irrelevant in the transaction, as the demand for the good at a particular price indicates that a consumer has made a capital expenditure on a machine that requires the source of energy and that he or she believes that in either the short or long term (a capital expenditure assumes the long term) the ready availability of the energy will continue to exist.

From a market transactions perspective, energy security issues are part of the transaction costs associated with energy consumption. The transactions costs are bundled into the retail price for the energy good, and are a measure of all transaction costs borne by all relevant parties associated with the recovery, processing, and delivery of the energy good. These costs include the costs to governments for defense expenditures associated with energy recovery, processing and delivery; it should be noted, of course, that the defense costs are also paid for by private citizens through their personal taxes.

With the exceptions of certain basic goods such as water and air, markets assume that nearly all other goods can be substituted theoretically for other goods. Therefore, as the price of certain energy goods increased due to rising transaction costs associated with production, processing, and transportation (all of which could be affected by rising costs of energy security), the demand for that good in relation to substitute goods would decline. Going back to the whale oil example for a moment, the demand for whale oil for lighting and other commercial purposes led to the overconsumption of whales. It was not long before the transaction costs associated with whale oil production rose substantially when the once plentiful marine mammals became relatively scarce. The time it required to hunt down whales increased as ships had to take extended cruises to the Arctic and Antarctic Oceans in search of whales and were finding whales which were smaller and of lower quality with respect to whale oil quantity. Longer cruises to more obscure locations increased the risk of piracy and threatened the whaling industry.

With the development of a domestic petroleum and natural gas industry, a chemically more powerful form of energy—one with substantially lower transaction costs—began to replace whale oil for use in lighting and in many other commercial uses. In today's world, petroleum faces the same challenges once faced by whalers in the 19th century. Production costs per barrel are rising because energy companies have to go to increasingly remote and, in many cases, dangerous (i.e., politically unstable or corruption-ridden) places to find petroleum. The days of "easy" petroleum are largely a thing of the past. The petroleum found now is often "heavy" and "sour," which means that the petroleum has a great deal of impurities and high concentration of sulfur, crude oil characteristics which require a great deal more energy (and higher cost) in the refining process.

Increased transaction costs and revolutionary technology breakthroughs have spurred the domestic petroleum industry to develop a shale oil industry. While it does not involve going to the ends of the Earth, shale oil does require going deep underground and involves the use of chemicals, water, and sand to loosen oil and natural gas from porous rock formations. Some types of the liquid fuel products recovered in the process could hardly be called "liquid"—shale oil recovered through this process cools to the consistency (but certainly not the purity) of the petroleum jelly found in the medicine cabinets of most homes.

Transaction costs in energy are often shaped by government regulation. In the 19th century, government regulation was fairly minimal. Therefore, the transaction costs associated with the movement from whale oil to kerosene and natural gas were not shaped to any significant degree by government regulatory policies. In the 20th and now 21st century, the regulatory situation has changed substantially. Government regulation impacts on energy today play a substantial role in shaping the transaction costs of energy supplies. Without repeating earlier points made in the chapter, much of this regulation relates to the relationship between elected officials, public sector bureaucrats, scientists, and public interest groups operating in the form of advocacy coalitions within policy subsystems articulating, through public policy regulating certain forms of energy, the marginal social costs of energy production, processing, transportation, and use.

The point here, however, is not to provide a dissertation on shale oil or whale oil. We seek simply to provide working examples of how transaction costs play a key role in the substitution of one energy commodity for another. In the case of energy, security can play a large role—in the case of petroleum, it most certainly does—in the transaction costs that shape wholesale and retail prices.

Increasingly consumers are restive and untrusting of rising energy prices. Adding to the restiveness is the fact that it is difficult to articulate the proportion of energy costs associated with energy security. In the case of privately traded commodities such as petroleum and natural gas, corporations are hesitant to publicly discuss the transaction costs associated with particular energy prices. In fact, it would be very difficult for them to do so because they do not set energy prices—it is a function of both supply and demand. On the supply side of things, transaction costs associated with energy security may mean that an energy-producing company will simply market a smaller quantity of liquid or gas fuels because it would not make economic sense to sell energy products at a price below the cost of production.

In the case of oil and gas production, quantity is limited not only by economic decisions about profit. Each oil and gas well has limits on productive capacity, which affects the variable costs. If the price of fossil energy is below the average variable cost, production would likely stop because the firm is unable to cover average costs. Costs of production would be impacted by energy security costs. As energy security costs rise, prices must rise in order to meet average costs of production. If prices do not rise, then the quantity produced will decline.

Thinking back to the definition of energy security outlined in the first chapter, readily available supplies of energy at an affordable price are considered to be key aspects of a "secure" energy source. From this perspective, energy security can be threatened by rising prices not only because of financial impact on individuals and firms, but also due to the unwillingness of oil producers to supply their goods at prices below rising costs. A demand for a secure and "cheap" supply of a resource may conflict directly with the market's ability to supply energy goods in a quantity and at price that consumers find attractive.

From an economic and scientific perspective, fossil energy resources remain cheap for what is gained from their use. Rising prices are, for the most part, a reflection of the increased energy security risks—the costs of securing the resources in increasingly remote and challenging locations and the supply chain costs of safely transporting the energy goods to distant markets—and the increased demand for fossil energy worldwide. Still, the prices are quite low, reflecting the relative glut of fossil energy. The world is still meeting a demand of over 90 million barrels per day at prices that are below 1980 peak oil prices. From a market perspective, fossil energy remains "secure" to the extent that supply is being met at prices that reflect growing costs of energy production and the increased costs of energy "security," to include the costs of environmental regulation.

As demand for fossil energy continues to grow and prices rise, substitutes for liquid fossil fuels—many of which are readily known to consumers—will become more attractive to consumers. A booming natural gas industry is already poised for increased demand for its products, readily quite available domestically and more "secure" in nearly all meanings of the concept, albeit it is not carbon neutral. What is particularly interesting is that a large proportion of shale gas and oil being produced in the United States is being developed on private lands; subsidies to industry largely occur in the form of transportation along public roads, and gas pipeline rights of way transecting public lands. Rail transportation occurs on privately owned rail lines.

Alternative and, as a subset, renewable energy sources are increasing in supply, most being heavily subsidized through public policy. In the modern era, nearly all forms of energy receive some form of incentive or subsidy that operates to encourage resource development; but alternative energy sources have received a substantial boost in recent years from government seeking to expand its energy portfolio. As a result, alternative energy sources have become price-competitive, which probably should not be the case given the public policy-influenced demand for such resources. The prices of alternative and renewable energy resources should be substantially higher given policy-influenced demand (e.g., state level renewable energy portfolio standards, state mandates for "greener" low-emission vehicles). Currently, alternative energy sources could not possibly meet the policy-created demands, and beyond that, could not possibly meet energy market demands—there is simply insufficient supply.

From another perspective, however, supply-related concerns are increasingly inducing commercial energy consumers to invest in energy alternatives. This is particularly true in electricity markets where business enterprises in the United States and elsewhere are betting that they can achieve greater energy security by reducing their exposure to grid-provided electricity. It is also true in the case of biofuels, where commercial interests are reducing exposure to fuel markets (particularly low-emission fuels demanded by government policy). Reflecting on the earlier observation that alternative energy is underpriced given rising demand, businesses seek to create their own secure energy supplies. Subsidies will come to an end eventually—assuming regulatory requirements for increased demand for renewable energy remain in place, the prices of renewable energy on a per unit basis will rise barring major breakthroughs in rare earth minerals demand for components, a revolution in manufacturing, and a fundamental change in millennia-old conceptualizations of property rights. Some of these requisite fundamental changes are already occurring and will, from a market perspective and otherwise, significantly affect energy security over the long term.

Conclusion

Government and markets tend to view energy security issues quite differently. Governments generally view energy security from the standpoint of the articulation of property rights, regulating those rights in ways that reduce identified marginal social costs, and ways that advance existing policy priorities. Government perspectives have evolved over time, reflecting the changing policy priorities of elected leaders, the precedential decisions of courts, ordinary citizens, advocacy groups and business interests. The establishment of property rights governing migratory sub-surface energy sources has relied heavily on the "rule of capture." In short, being first to a location and using your ingenuity to access a resource are the basic principles governing property rights to the extraction of sub-surface migratory energy resources. Such legal principles, however, are not uniformly applied to other needed migratory sub-surface resources, such as water.

Marginal social costs prove to be the basis of regulating property and its use. Environmental costs associated with mineral resource extraction from public lands, for instance, are identifiable marginal social costs that may serve as the basis of energy policy shifts—that is, identifying the true costs of fossil energy and, through regulation, pushing up the average costs of producing domestic fossil energy. But, does this effort to manage marginal social costs through constraints on property rights have a long-term impact on energy security? From the perspective of those concerned with climate change, the answer is a distinct "yes," but with the caveat that other replacement energy sources are capable of meeting demand and doing so in a way that is affordable.[31]

In the promotion of energy alternatives, public policy and the marketplace have witnessed significant changes in the way that property rights are conceived. Property rights of owners on lands adjacent to renewable energy users are impinged upon by the higher priority offered to the property rights of the latter group. Water rights needed to operate large scale renewable energy facilities on public lands may receive the benefits offered by access to Federal reserved water rights, particularly in cases where renewable energy facilities are located on Native American reservations.

From a market perspective, energy type, the uses made of the energy, and the development decisions regarding resource recovery are based on a combination of market demand, the ability of firms to minimize transaction costs, and profitability. If costs of production exceed market price, then the long-term production and sale of a particular form of energy is unlikely to occur. Transaction costs are managed either through government policy or through the market mechanism. For instance, energy security poses transaction costs that are in some cases managed through the use of private sector security firms. In other cases, such as electricity generation and transmission, government regulation of electricity markets serves as a basis for promoting energy security.

In the United States, government-owned or semi-public corporations play a smaller and oftentimes slightly different role in energy production and distribution than is the case in other nations. The government-owned or semi-public corporations play more noticeable roles in nuclear and electricity markets than in the still-dominant fossil energy market. In the case of any government-owned or semi-public enterprise, sunk costs are more likely to be absorbed by governments through budget and finance decisions, facilitating more rapid shifts in priorities. To a much greater extent than government or quasi-government entities, private publically-traded energy corporations must consider the role of sunk costs in their decision-making. Amortization of capital goods associated with production is an important part of remaining profitable, the latter being of lesser importance in the case of government-owned or semi-public corporations. Therefore, rapid changes in energy markets have an impact on property rights of energy firms which have made financial decisions on the basis of certain market rules and incentives emerging from public policy. In the case of private renewable energy firms, of course, equipment is new, in some ways subsidized by government policy and incentives. For private renewable energy firms, however, the lesson to take away from history is that times change and policy priorities evolve—in earlier decades, the fossil energy firms received similar governmental largesse as do contemporary renewable energy firms. The bigger lesson for all those concerned with energy security matters is that energy security is a function of the allocation and articulation of property rights within our legal institutional framework. Property rights and the exercise of such rights are key components in long-term market developments in energy supply, affordability, and long-term security. Shifts in property right articulation have real impacts on markets and real impacts on energy security. In the case of nations with large national energy agencies or corporations managing key energy resources, the capacity to create a more uniform articulation of property rights and rules governing property use may lead to greater capacity to promote energy security. Conversely, large-scale government control has the potential to reduce the entry points of energy innovation, as witnessed in the case of Pemex in neighboring Mexico.[32] Perhaps even more pointedly, witness the energy supply revolution occurring on private lands in the United States, recent developments which have led the nation to becoming the largest producer of petroleum on a daily basis. Market-driven innovation often works in this highly dramatic way, but property rights must be carefully and uniformly articulated, and market incentives must be actively considered and incorporated into an ever-evolving energy security calculus.

Notes

1 Ronald Inglehart and Paul R. Abramson, *Value Change in Global Perspective*, Ann Arbor, MI: University of Michigan Press, 1995; Ronald Inglehart, *Culture Shift in Advanced Industrial Society*, Princeton, NJ: Princeton University Press, 1990.
2 *Acton v. Blundell*, 152 Eng. Rep. 1223, 1235, 1843.
3 David E. Pierce, "Coordinated Reservoir Development: An Alternative to the Rule of Capture for the Ownership and Development of Oil and Gas," *The Journal of Energy Law and Policy*, 4(1), 1983, pp. 1–79.
4 Stephen D. Alfers, "American Mining Law Reform," *Journal of Energy and Natural Resources Law*, 12, 1994, pp. 424–441; Mark Squillace, "The Enduring Vitality of the General Mining Law of 1872," *Environmental Law Reporter*, 18(July), 1988, pp. 10261–10270.
5 Marc Humphries, "Mining on Federal Lands: Hardrock Minerals," *National Council on Science and the Environment*, November 6, 2007, p. 1; see also *U.S. v. Midwest Oil Co.* 236 US 459 (1915) in which the Court upheld the power of the Executive to use Presidential powers to withdrawal public lands from energy development.
6 US Department of the Interior, *Oil and Gas Lease Utilization, Onshore and Offshore*, Washington, DC: DOI, 2012, www.blm.gov/pgdata/etc/medialib/blm/wy/programs/energy/og/leasing/protests/2012/may/Appeals.Par.42771.File.dat/ExhP.pdf (accessed August 1, 2013).
7 John Filostrat, "Obama Administration Announces 21 Million Acre Oil and Gas Lease Sale Offshore Texas: August Auction to Offer All Unleased Acreage in Western Gulf of Mexico," *U.S. Bureau of Interior*, April 24. www.doi.gov/news/pressreleases/obama-administration-announces-21-million-acre-oil-and-gas-lease-sale-offshore-texas.cfm (accessed August 1, 2013).
8 Pew Research Center, *Little Change in Opinions about Global Warming: Increased Partisan Divide on Energy Issues*, Washington, DC: Pew Research Center, 2010; Theodore J. Lowi, "Four Systems of Policy, Politics, and Choice," *Public Administration Review*, 32(4), 1997, pp. 298–310.
9 Richard Nemec, "Taming the Deregulation Beast: After a Hundred Years of Official Monopoly, Government-run Utilities Will Soon Enter the Free Market; Will the Utilities Adapt and Survive, or Will They Be Trampled by the Private-sector Giants?" *California Journal*, 28, 1997, pp. 24–26.
10 John R. Munkirs, Michael Ayers, and Al Grandys, "Rape of the Rate Payer: Monopoly Overcharges in the Regulated Electric-Utility Industry," *Antitrust Law and Economic Review*, 57, 1976–1977, pp. 57–68.
11 Anthony Sampson, *The Seven Sisters: The Great Oil Companies and the World They Made*. New York: Bantam, 1976.
12 John V. Mitchell, and Beth Mitchell, "Structural Crisis in the Oil and Gas Industry," *Energy Policy*, 64, 2014, pp. 36–42.
13 Inho Kim, "Who Bears the Lion's Share of a Black Pie of Oil Pollution Costs?" *Ocean Development and International Law*, 41(1), 2010, pp. 55–76.
14 Cliff Rochlin, "Is Electricity a Right?" *The Electricity Journal*, 15(2), 2002, pp. 31–36.
15 Michael Grubb, *Emerging Energy Technologies: Impacts and Policy Implications*, Farnham: Ashgate Publishing, 1992.
16 John Rawls, *A Theory of Justice*, Cambridge, MA: Harvard University Press, 1971.

17 Z.M. Chen and G.Q. Chen, "An Overview of Energy Consumption of the Globalized World Economy," *Energy Policy*, 39(10), 2011, pp. 5920–5928.

18 Jane Adams, "Oil Shale Boom Creates Many Issues," *The Southern Illinoisan*, March 19, 2013, http://thesouthern.com/news/opinion/editorial/guest/oil-shale-boom-creates-ma ny-issues/article_87c6e8c4-9053-11e2-bbb1-001a4bcf887a.html (accessed August 3, 2013). In the report *Boom and Bust in the American West*, the Center of the American West highlights the impact of business cycles on local and state economies in the American West. National government policy changes have impacts on the boom/bust local and state economies. For example, the end to the Cold War and base realignment led to the closure of the biggest industry in many local towns—that is, the loss of bases and base personnel who dwelled in towns and made many purchases of homes, furnishings, food and clothing. The paper points out that this boom/bust cycle is not atypical for communities with highly concentrated economic sectors that are heavily dependent on single industries to provide for the local tax base and salaries needed to fuel local economies. See Patricia N. Limerick, et al., *Boom Bust in the American West*, Boulder, CO: Center for the American West, www.centerwest.org/publications/pdf/ boombust.pdf (accessed August 15, 2014).

19 Kenneth J. Arrow, "A Difficulty in the Concept of Social Welfare," *Journal of Political Economy*, 58(4), 1950, pp. 328–346.

20 Aaron B. Wildavsky, *Speaking Truth to Power: The Art and Craft of Policy Analysis*, Boston, MA: Little, Brown, 1979.

21 47 U.S. 344 (1848).

22 David H. Getches, "Managing the Public Lands: The Authority of the Executive to Withdraw Lands," *Natural Resources Journal*, 22, 1982, 279–335.

23 Carrie Covington Doyle, "Note & Comment: The Modern Oil Shale Boom: An Opportunity for Thoughtful Mineral Development," *Colorado Journal of International Environmental Law and Policy*, 20, 2009, p. 260.

24 Bureau of Land Management (BLM), *Mineral Leasing Act of 1920, as Amended (8/9/2007)*. Washington, DC: Bureau of Land Management, 2007, www.blm.gov/pgdata/etc/media lib/blm/ut/vernal_fo/lands___minerals.Par.6287.File.dat/MineralLeasingAct1920.pdf (accessed August 15, 2013).

25 BLM, *Mineral Leasing Act*, pp. 3, 104.

26 BLM, *Mineral Leasing Act*, p. 90.

27 Thomas C. Campbell, "Allocating a Scarce Resource," *American Water Works Journal*, 77(9), 1985, p. 54.

28 Debbie Leonard, "Doctrinal Uncertainty in the Law of Federal Reserved Water Rights: The Potential Impact on Renewable Energy Development," *Natural Resources Journal*, 50, 2010, pp. 611–643.

29 Joseph Stewart, Jr., David M. Hedge, and James P. Lester, *Public Policy: An Evolutionary Approach*, Stamford, CT: Cengage Learning, 2007.

30 Chennat Gopalakrishnan, "The Doctrine of Prior Appropriation and Its Impact on Water Development: A Critical Survey," *American Journal of Economics and Sociology*, 32(1), 1973, pp. 61–72.

31 Anthony D. Owen, "Environmental Externalities, Market Distortions, and the Economics of Renewable Energy Technologies," *Energy Journal*, 25(3), 2004, pp. 127–156.

32 Adam Williams, "Pemex, Mexico's State Oil Giant, Braces for the Country's New Energy Landscape," *The Washington Post*, June 7, 2014, www.washingtonpost.com/ business/pemex-mexicos-state-oil-giant-braces-for-a-the-countrys-new-energy-landscap e/2014/06/04/07d171d6-ea69-11e3-93d2-edd4be1f5d9e_story.html (accessed July 4, 2014).

4

DOMESTIC POLICY AND ENERGY TRANSFORMATION

Fossil fuels

The pursuit of energy security involves multiple interactions among global commodity markets, production technologies, public policies and consumers, and it is characterized by ground that is continually shifting. This state of flux seems particularly interesting in the context of rapid growth in the production of US domestic energy sources, a development that was largely unanticipated only a few years before it happened. This expansion of domestic energy production includes not only oil and natural gas, made possible by technological innovation involving hydraulic fracturing and horizontal drilling, but also solar and wind power generation, which have seen falling costs along with high rates of industry growth and consumer adoption. Each of these parallel energy industry developments, characterized by Michael Levi in *The Power Surge* as unfolding revolutions in the energy sector, suggests that the goal of achieving greater energy security is now more attainable.[1] These developments exemplify a transformation in the energy sector (or a series of transformations) that have been, at the same time, both actively sought in an effort to increase energy security, and largely unplanned, leading to both favorable and unfavorable unanticipated consequences.

These changes in the energy sector are, of course, not only a function of energy markets and associated production and commercialization technologies, but of public policies that have simultaneously responded to these markets and technologies, and that have also been designed with the aim of building or altering markets for renewables and fossil fuels by encouraging the adoption of particular technologies by industries and consumers. This is what it means for us to understand the status of energy as a "marketable public good," in which public policy is used to set the parameters of transactions in private markets involving what is in part a public good, all to provide social benefits not offered by relying on purely private market transactions.

In consideration of the energy concerns of the past couple of
have featured high oil prices, dependency on foreign oil, high leve.
inefficiency in almost all forms of energy use, wars in the Persian Gu.
Afghanistan, and global climate change—the pressure for policy change has been
significant. Major energy legislation at the federal level was enacted in 1992,
2005, 2007 and again in 2009, along with numerous energy-related measures that
were more limited in scope. At the same time, many US states have pursued their
own energy policy initiatives to move beyond what the federal government has
done in the energy policy area. The broad aim of this national and multi-state
effort has been a transformation of the energy sector that is aimed at diminishing
the adverse economic and environmental effects of meeting the nation's energy
needs, while providing more abundant, affordable, reliable, clean, diversified
supplies of energy to commercial and residential consumers.

This transformation is being sought with regard to both traditional fossil fuels
and renewable energy sources. To elucidate this understanding of such a transformation, we analyze four different issues and the policy environments that
govern them. These are the four issues that most fully exemplify this transformational effort—namely, the *shale oil and gas revolution*, the *regulation of coal use in
electricity generation*, the *integration of wind and solar power into the electricity supply*, and
the effort to increase the use of *biofuels*. These issues have come to take on
increasing importance in the debate over energy, energy security, and energy
policy, and they serve to illustrate how public policy affects (both positively and
negatively) the achievement of domestic energy security goals in the United States.

The Fracking Revolution

The greatest change to the American energy landscape in recent years, one that
seemed to come virtually out of nowhere and surprise almost everyone associated
with the energy industry, has been the rapid growth of large-scale shale oil and
natural gas production. As Michael Levi aptly points out in *The Power Surge*, "as
recently as 2009, you couldn't even find the words 'shale gas' in the annual U.S.
government energy outlook."[2] The change began in the natural gas industry first.
Energy forecasters in the United States had been expecting natural gas production
to decline, and the industry was preparing for greater levels of natural gas
imports.[3] In fact, prior to 2010 efforts were started in the United States to build
facilities that would import liquefied natural gas (LNG). Natural gas prices also
rose and fell repeatedly (from a low of $1.34 per million BTU in 1998 to a high
of $14.49 per million BTU in 2005) in what had historically been a volatile
market.[4]

The usage of two production technologies in combination brought about a
dramatic change: horizontal drilling and hydraulic fracturing, or "fracking" in lay
terms.[5] After drilling down to the level where the oil and gas lies—anywhere
from 1,000 feet to a couple of miles—workers drill horizontally for up to several

thousand feet. A mixture of water, sand and chemicals are then injected into the opening at high pressure and explosive charges are set off deep underground. This combination of pressure, materials injection and explosive charges creates fissures in the shale rock, making the oil and gas trapped within accessible for bringing to the surface. While both of these techniques had been developed long ago and were in limited use for decades, they were not considered to be very useful in the oil and gas industry for accessing hydrocarbon reserves. It was the persistence of George Mitchell at Mitchell Energy and Development that led to the engineering and commercial breakthrough in the fracking arena in the late 1990s. The company was looking for ways to access large volumes of gas in shale rock, and after a great deal of experimentation the engineers at Mitchell Energy and Development found the optimal formula for the injection materials and the sequencing of the process. After several years, as other firms began to use the same process to access natural gas (including Devon Energy, which purchased Mitchell Energy in 2002 for $3.5 billion), the industrial process breakthrough led to the current boom, and the process was also applied to accessing "tight" oil trapped in shale formations.[6]

This new ability to tap into previously unavailable reserves of shale oil and gas has led to the remarkable growth of domestic production in recent years. New shale oil and gas wells in the United States now number in the thousands.[7] Only as recently as 2005 the United States imported two-thirds of its daily oil needs. In 2014 and 2015, the figure moved to 40%, with North Dakota and Texas leading the way with the greatest enhancement of output.[8] The Bakken Shale in North Dakota propelled the state past Alaska as the country's second largest oil-producing state, producing just over 38 million barrels of oil a month at the end of 2014, a record high. Only two years earlier, the figure was just under 17 million barrels per month, and 7.3 million two years before that. In the wake of a drop in oil prices, production remained high at 37 million barrels per month in July 2015.[9] In Texas, production in the Permian Basin and the Eagle Ford Shale has soared, leading the state to produce over 100 million barrels of oil per month in early 2015, roughly 35% of all US oil production.[10] In both US states new wells continually came into active production for several years, but there was a rapid reduction in drilling new wells as oil prices fell rapidly in 2014 and 2015. According to projections from both the US Energy Information Administration and the International Energy Agency, in projections made before the market slowdown, the United States was expected to pass Saudi Arabia as the world's largest oil producer in 2015, though it is not expected to hold this position for many years.[11] In fact, the US reached that mark in December 2014 with a high of 12.4 million barrels per day of oil and natural gas liquids.[12]

It is important to point out that this was wholly unexpected just a few years before the boom started. The extent to which this level of production elicited continual surprise, even among industry experts, was clearly captured in a Bloomberg news report issued in early 2014, which quoted an analyst at a British energy consultancy as saying, "I don't really think anyone saw this coming … The U.S.

shale boom happened much faster than people thought. We're in the middle of a new game. There's nothing in the past that predicts what the future will be."[13]

With regard to shale gas in the United States, the story is a similar one to what has occurred in domestic petroleum production. Shale gas provided 40% of the country's gas in 2014, roughly 12 trillion cubic feet out of 31 trillion cubic feet (TCF).[14] The Marcellus Shale, which stretches from West Virginia through Ohio and Pennsylvania into New York, provides the largest source of shale gas; it has led the way in the United States with regard to shale gas production. Even as many other sites in the United States saw no or little growth in 2014, gas output from the Marcellus Shale continued to rise.[15] This development has major implications for the nation's energy profile, and its energy security in general. While the United States remains a net importer of natural gas, the level of net imports has been cut by more than half since 2008, from 3 TCF to 1.2 TCF in 2014. And the United States is expected to become a net exporter of natural gas by 2020.[16] This is truly a dramatic turn from only a few years ago, when talk of natural gas supplies in the country was about building terminals to import liquefied natural gas at strategic port sites in the nation. All this production has caused the price of natural gas to fall greatly and stay at a low level, as Figure 4.1 shows. In 2014, the Henry Hub spot price dropped as low as $3.26 per million BTU (from a high in 2008 of $13.20), and even that was an increase over the low of $1.86 in 2012.[17]

This revolution in domestic oil and gas production has had numerous impacts—ranging from long-term shifts in global energy markets to the "boom-town" effects (and in the oil fields, the "bust" effects) in communities where unconventional oil and gas production occurs. With respect to the aim of enhancing energy security, the increased production of domestic shale oil and gas most certainly advances the goals of increasing supplies, strengthening the reliability of the fuel supply chain, reducing costs, keeping more energy dollars at home, and diversifying the point of origin of energy sources. It also seems likely

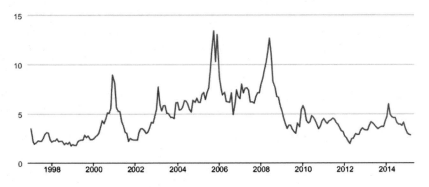

FIGURE 4.1 Henry Hub Natural Gas Spot Price (in dollars)
Source: US Energy Information Administration.

to mitigate some major national security vulnerabilities, though perhaps not in the short run. In the case of natural gas, its use could also potentially offer an environmental benefit if it is used in place of coal to generate electricity, though there is much controversy about the environmental impacts of fracking, even as a replacement for coal. Still, there is often little agreement regarding the benefits and costs of these impacts given how recent these dramatic developments have been on the energy supply side.

US National Security

One of the most significant changes resulting from the growth of domestic oil and gas production has been the understanding, particularly within the United States, that US national security, along with global security, are likely to be impacted in both significant and positive ways. These implications are addressed in Chapter 6, which examines the global dimensions of US energy security and policy. Suffice it to say here that if and when policymakers, business leaders and citizens come to increasingly think of the United States as an energy exporter, or at least as less of a consumer/importer of foreign energy supplies, then the debate over energy security, energy policy, and US national security will change greatly, as it already has begun to do.

Economic Development, or a Gold Rush Boom and Bust Cycle?

The shale oil and gas revolution has had an enormous positive economic impact, both on the communities and regions where production occurs and on the nation as a whole. To name but a few effects:

- New, high-paying jobs in oil and gas fields.[18]
- Growth of employment and incomes in businesses supplying the oil and gas industry—chemicals, vehicles, drilling equipment, sand, etc.
- New jobs and incomes in local businesses serving oil and gas workers—restaurants, housing rentals, hotels, clothing and shoe stores, big box retailers, personal services providers, etc.
- Reduced cost of energy inputs for industry (steel, cement, glass, aluminum, automobiles) and electricity generation (including the displacement of coal, nuclear and wind power).
- Reduced trade deficit from fewer oil and gas imports.

Several studies have been undertaken to estimate the full economic impact. The consultancy IHS, for example, released a series of detailed reports on the subject. One report issued in 2011 stated that 150,000 people in the United States were employed in the shale gas industry in 2010, and that this figure would increase to 250,000 in ten years. With regard to indirect employment (suppliers to the industry) and induced employment (the local businesses doing well due to all the new income being spent by people in the area), IHS estimated that another 450,000 jobs had been added to the economy due to shale gas. By 2020, the researchers at IHS estimated that this economic activity could add $150 billion to the American economy.[19] Subsequent studies, comprising a three-part series, reported that in 2012, shale oil and gas supported 2.1 million jobs, provided $74 billion in federal and state revenues, added $283 billion to the gross domestic product, and increased household income by an average of $1,200.[20] By 2025, IHS concluded, unconventional oil and gas would account for more than $530 billion in economic activity and provide 3.8 million jobs in the US.[21]

By the same token, a study completed in 2013 by PricewaterhouseCoopers (PwC) for the American Petroleum Institute looked at the economic impact of all oil and natural gas operations in the United States (not just shale oil and gas). It found that the industry accounted for 9.8 million full-time or part-time jobs, which was more than 5% of the country's total employment. It also concluded that oil and gas operations and investments comprised $1.2 trillion, a full 8% of the American economy.[22]

A corollary to this assessment on the economic impacts of shale oil and gas is that the availability of plentiful, cheap energy is fueling a "manufacturing renaissance" in the United States. A number of studies have attempted to make this case in the past few years.[23] The case for this, as one report released by the US Conference of Mayors points out, is that unconventional shale oil and gas, by offering accessible and inexpensive energy resources, have boosted growth in both the manufacturing sector and the national economy across the board. In particular, demand for new pipelines and drilling equipment sparked growth in steel, iron, fabricated metals, and manufactured machinery. At the same time, the advent of relatively cheap domestic energy supplies supported the expansion and productivity of manufacturing plastic, rubber, chemicals, cement, and glass, among other commodities.[24] These beneficial economic impacts can be expected to be long-lasting, fueling long-term growth of the economy and offering US manufacturers a competitive advantage in the global marketplace.

Overall, these studies suggest that the American economy owes a great deal of its success to the oil and gas industries. However, the generally accepted view that the advent of major new sources of oil and gas represents a long-term, structural benefit to the American economy is not universally accepted. There are three types of arguments offered to call this assertion into question. The first is that if cheap energy were the key to economic success, one should expect that Iran, Venezuela, Nigeria and other countries rich in energy resources would be much stronger economically than they are. It is argued that what has happened in the United States with respect to the recovery from the "Big Recession" is less a renaissance fueled by plentiful, cheap energy than it is a recovery of an advanced economy that had suffered a particularly severe cyclical downturn. It is argued further that the nation's demonstrated ability to be increasingly productive results from a complex combination of the business cycle, infrastructure developments, a legal and financial system that promotes and protects investment and entrepreneurship, a skilled labor force, and a continental consumer base of great breadth. Inexpensive energy is clearly an added benefit, but it can plausibly be argued that abundant, low-priced energy hardly represents the key driver of economic vitality.

A second argument suggests that the economic impact has indeed been real—one needs only to visit towns near oil and gas operations in North Dakota and Texas to see the impact of the enormous influx of jobs, money and people in these communities. A news report from 2011 pointed out that in the Williston Basin of North Dakota, "you can make $15 an hour serving tacos, $25 an hour waiting tables and $80,000 a year driving trucks."[25] However, it is suggested that the boom is more like a gold rush in local economies than a long-term structural change to the US economy. At the local level economies have been upended, as the price for just about everything goes up, but only as long as the wells can keep producing. Moreover, the impacts of rapid development on local communities can be quite uneven, as long-time residents find themselves priced out of the housing market, and something as simple as a night out at a restaurant can cost two to three times as much as it did in the recent past. This "boomtown" effect has been a common phenomenon associated with earlier natural resource discovery and development episodes in our nation's past, and the concerns are similar today.[26]

Prior to the rapid fall in oil prices, the economic concern was that once the marginal production became too expensive, the energy companies, the field production workers and the money would be expected to disappear, turning the boom into a bust if great care is not taken to provide for the simultaneous development of alternative sources of economic activity. Since the output of wells is expected to decrease quickly, new wells must continually be found relatively nearby to maintain ambitious rates of growth in production, and the likelihood of this occurring over the long run is small. As Herb Stein, former Chair of the Council of Economic Advisors put it, "If something cannot go on forever, it will stop."[27] Even the optimists looking at the US surpassing Saudi Arabia in oil production are aware that this will be a temporary phenomenon, with clear limits to be experienced on continued growth in new exploitable sources. Since the slowdown in the oil sector starting in 2014, the drop in prices has had a rippling effect throughout the local economies in places like North Dakota and Texas, and it came much sooner than anyone expected since output at wells decreased not because of geophysical characteristics, but because of market changes. The workers and communities that had previously benefitted from high oil prices began to find themselves suffering in what has been increasingly called a "bust" in the oil economy.[28] Since oil at $50 a barrel may not be worth getting out of the ground, oil companies have slowed or stopped drilling new wells (though production at existing wells has largely continued). Employees at the oil companies and their subcontractors experienced layoffs and found that jobs at new wells were few and far between. The expectations of future work and high incomes abruptly changed.

The third argument made by the critics of the affordable energy-based recovery argument is that the economic benefits of the oil and gas boom tend to be greatly overstated. This claim is based on the idea that business expenditures on energy, particularly natural gas, are within historical norms, and that as a proportion of total manufacturing costs, energy is not a major expense for most US domestic industries.[29] This critique is also grounded in an understanding that the various economic studies tend to employ methodologies that seek to account for all types of economic activity associated with energy development and its downstream impacts. Therefore, they have a broad reach to account for transactions throughout the economy, both directly and indirectly related to the oil and gas industry. This makes the numbers reported in the studies cited above very large, particularly with regard to the projections about future expectations. Such optimistic scenarios can elicit skepticism regarding the favorable conclusions reached. At the same time, like most studies that seek to project economic activity into the future, there is a tendency to assume that conditions in the present will largely be replicated in the future. In the unpredictable world of the oil and gas sectors, forecasting energy prices is notoriously difficult. (As the great Yogi Berra said, "It's tough to make predictions, especially about the future.") The studies cited above were all completed prior to the drop in oil prices in 2014, and therefore, the expectations regarding job creation and economic growth in the US require revision. For example, inexpensive oil will over a period of time reduce the number of jobs in the oil fields, but it will be likely to also allow for increased spending by consumers in other sectors as their energy bills drop. The point here is not to criticize such studies as unhelpful or unreliable. Rather, it is to point out the importance of recognizing their assumptions in assessing the potential of the oil and gas industries. It is also important to be cognizant of the fact that even if the numbers of jobs or dollars resulting from the fracking boom may be overstated, and even acknowledging an oil market slowdown, it is clear that the economic impact of the growth of this sector has been and can continue to be significant by any reasonable measure.

An additional critique of the purported economic benefits of the shale oil and gas revolution is that there are relatively high environmental costs resulting from unconventional oil and gas production that are not accounted for in the calculations made of economic benefits. It is argued that when one considers these "externalized costs" appropriately, the claimed economic and societal benefits of cheap, plentiful energy are considerably diminished. These environmental considerations are addressed in the following section.

The Environmental Impacts

It is likely that a generation ago, in the wake of the oil shocks of the 1970s, the use of fracking to boost domestic supplies of oil and natural gas would not have generated the level of opposition over environmental impacts that it has given rise to today. After all, the primary aims of US energy policy have been sufficient supplies at affordable prices, and shale oil and gas offer these benefits to great effect. However, greater attention to sustainability means that the amazing growth of domestic oil and gas supplies is not universally considered to be a benefit.

The acquisition of oil and gas from hydraulic fracturing includes a wide range of production site activities. Each well requires a sizable industrial endeavor. The process involves drilling deep underground, often through water supplies that serve local communities, injecting millions of gallons of water, tens of thousands of gallons of chemicals, and tons of sand into the ground. It also entails setting off blast charges to open up fissures in the shale rock deep underground, and ultimately extracting oil or gas from the well site. This set of activities constitutes only part of the process, which also includes clearing land for well pads and associated operations; constructing access roads and other associated infrastructure such as pipelines; managing and storing the flowback water and sludge that comes out of the ground when the drilling process is taking place; transporting and processing the fossil fuels that are extracted from under the ground; and transporting millions of gallons of water and wastewater for use, treatment, and/or disposal in tankers and storage containers. When one multiplies this times the thousands of wells in the United States, and the fact that companies often will drill multiple, closely-spaced vertical wells at a single site to maximize access to the oil and gas, the scope of the larger environmental impacts can rightly be understood as quite significant.[30] It is, in fact, the known environmental impacts and the yet unknown effects upon groundwater supplies and geological formations such as faults that have prompted the greatest concern and opposition to fracking in the United States and elsewhere in the world.[31] The two most prominent concerns commonly voiced by opponents to unconventional oil and gas production, such as the Natural Resource Defense Council and the Union of Concerned Scientists, deal with the public health and environmental effects of hydraulic fracturing on domestic water supplies and agricultural water quality (this includes the migration of both the gas and the fracking chemicals), and the net effect on greenhouse gas emissions. Other issues that have received less attention include fracking-induced earthquakes and air pollution from particulate matter, carbon monoxide, nitrogen oxides and volatile organic compounds released into the environment in the course of drilling operations and fossil fuel recovery.

There are a few key concerns with regard to water resources: the amount that needs to be used, the volume and proper disposal/storage of liquid wastes, and the contamination of groundwater and drinking water for residents in the area of fracking operations are all matters of concern. With respect to water usage, oil and gas production requires large amounts. Each well requires anywhere from 2–5 million gallons of water, a figure generally agreed upon by both industry and researchers.[32] At the higher end, for example, in Pennsylvania and West Virginia, where natural gas from the Marcellus Shale is produced, each well requires anywhere from four to five million gallons of fluid for injection.[33] The vast majority of this fluid consists of water, while a small portion consists of the chemicals injected into the ground, and these are mixed with sand. In West Virginia, over 80% of the water used is surface water from nearby rivers and streams, which can have an effect on local supplies for agriculture, industry and residential usage, especially depending upon the time of year when withdrawals are made.[34] In more arid places, such as Texas, water supplies are often shipped to the drilling sites over long distances. Nationwide, this translates into billions of gallons per year for carrying out fracking operations. However, compared to other forms of energy, a resource such as shale gas appears to be less water-intensive than other sources when one considers how much water is used to produce or acquire energy resources such as coal and ethanol derived from planted crops. A study from Harvard's Belfer Center found that the extraction and processing of shale gas uses 0.6 to 1.8 gallons of water per million BTUs (MMBtu), while onshore oil production uses anywhere from 1–62 gallons per MMBtu, depending upon the techniques used. Ethanol can require 100 gallons per MMBtu at the low end, and more than 3,000 gallons (to grow corn and process it into fuel) at the high end.[35] The energy industry and supporters of shale oil and gas production often point out that "many other activities—even those far less important than energy production, such as watering residential lawns and maintaining golf courses—use substantially more water than fracking."[36] However, these practices are far less controversial.

While the amount of water required is a concern, the bigger concern involving water is water quality, and the impacts that hydraulic fracturing can have on the contamination of groundwater and surface water due to the chemical-filled wastewater that is produced in the course of energy production. Also of concern is the problem of leakage of natural gas into water supplies. Some of the water, chemicals and sand mix used in fracking does remain in the ground, so it needs to be kept clear of municipal water supplies. Similarly, the millions of gallons that return to the surface have to be managed properly by well site operators. What both of these things mean in practice, and the extent to which they do occur on a regular basis in fracking operations, along with the possible adverse impacts of exposure to the wastewater, are the subject of much dispute and controversy in both the research literature and the advocacy-group studies disseminated to persuade citizens about the benefits and the possible downside of fracking.

The potentially harmful effects of unconventional oil and gas production on local domestic water supplies were thrust into the public imagination with the publication of a widely read story in *The New York Times Magazine* in 2011.[37] The report talked about how people living near fracking operations and wastewater impoundment pools filled with the chemically treated water used in the process were getting sick, and simultaneously finding dangerously high levels of benzene and arsenic in their blood; of how the air smelled like "sweet metal, rotten eggs and raw sewage"; how sometimes the water was black and ate away at faucets, hot water heaters and dishwashers; and how companies did not have to disclose the chemicals they use because that information is considered to be privileged as commercial proprietary secrets. The article prompted a number of critiques, but in spite of them, the article exemplified the growing sense, contrary to the oil and gas industry's reassurances, that hydraulic fracturing can be dangerous to public health and harmful to the environment. In addition, the article revealed the surprise to many readers that the government was allowing this to happen with unusual exemptions from the normal requirements of public disclosure of the presence of toxic materials normally provided to state and local emergency management and first responder officials.

The extent to which oil and gas drilling operations are the cause of water pollution has become the subject of a great deal of recent research and investigation, and remains ongoing. Academic, private and government studies and reports routinely feature the subject. A review of the scientific literature in 2015 by Adam Reed found that, "well casing leaks have been implicated in methane-contaminated drinking water in water wells overlaying the Marcellus and Barnett shale formations," and that "[c]ontamination via wastewater is also found to be a risk that has led to adverse impacts in the Ohio River valley." It also cited studies that showed how "[c]omputer modeling of contamination pathways, combined with probability analysis, indicate the scope and level at which contamination can occur through fissures in the shale rock."[38] An EPA report from 2004 on the subject concluded that hydraulic fracturing was not harmful to water supplies, but these conclusions were repeatedly challenged in the years after the study was conducted, particularly by numerous US environmental organizations.[39] In large part, studies addressing the impacts of fracking on water quality suggest that natural gas and flowback water sometimes do mix with surrounding water supplies. Numerous pathways for contamination have been identified, and the harmful impacts of the chemicals most likely being used are now relatively well known. What is becoming clear is that the potential for large-scale contamination of surface and groundwater exists, and becomes greater as oil and gas production from hydraulic fracturing in the United States grows in scale. What is less clear is the extent to which such environmental harm is actually occurring, and how public health is being affected by this type of fossil fuel extraction practice. As Reed noted, "most of the authors who studied this topic seem to agree that there is little evidence that hydraulic fracturing is directly contaminating water sources via migration of gas or chemicals through fissures. These findings indicate either that this pathway of water contamination poses minimal risk, or that insufficient data are available to make a clear determination."[40] Amidst such uncertainty, calls for additional study, testing and systematic monitoring richly populate the literature, pointing out that not enough has been done to fully determine the scope of the problem or document the effects, both short- and long-term. The need for additional research has prompted a new comprehensive effort by the EPA to study the impact of fracking on drinking water sources, looking at existing data from oil and gas companies, the scientific literature, watershed scale computer modeling, laboratory tests, and several noteworthy case studies. However, the results of the study at this writing (late 2015) have not been released and remain eagerly anticipated by proponents and opponents alike.[41]

In addition to concerns over water supplies, the impact from hydraulic fracturing on greenhouse gas emissions represents another major element of concern in some circles, but also of optimism for some others. The impacts can be quite difficult to assess, as there are a number of variables to consider in such estimations of impact. The argument largely revolves around shale gas, and not so much oil. On the positive side, with respect to shale gas, when the fuel is used for electricity production, and new gas-fired plants are being brought into service and replacing coal-fired electricity, the beneficial results are quite unambiguous. Coal is the dirtiest of fossil fuels, with harmful environmental impacts occurring throughout the mining, transportation and combustion processes. In addition, carbon dioxide emissions from the burning of coal exceed those from all other forms of energy. Considering that the proportion of electricity generation from coal, as well as the total number of kilowatt-hours, have declined, while these numbers have risen for natural gas, there is a compelling case to be made that natural gas is providing a noteworthy beneficial effect on US greenhouse gas emissions. Moreover, the argument for shale gas includes the fact that natural gas plants complement the use of wind and solar technologies better than do coal plants (gas-fired plants can be cycled on quickly as variable wind or solar electricity generation fluctuates), and that "coal mining removes entire mountains and contaminates streams with hazardous waste, [while] natural gas drill pads occupy only a few hundred square feet, and there are only a handful of cases of groundwater contamination by fracking chemicals."[42] These claims hold merit and make a strong case for the use of fracking to produce natural gas. At the same time, the critique of unconventional gas production in particular addresses a different issue altogether in terms of carbon emissions—namely, the extent of "methane leakage" that occurs in the fracking process. Since methane (unburned natural gas) is a far more potent greenhouse gas than carbon dioxide, significant leakage at drilling sites could potentially offset any gains made by using natural gas instead of coal for the production of electrical power. One widely cited (though controversial) study concluded that if natural gas leaks amounted to more than 3.2 percent of the overall production, then the usage of fracking-derived natural gas would be worse than continuing to use coal-fired power plants.[43]

Several studies have sought to determine the precise amount of leakage occurring, employing different methods (for example, testing samples at the drilling site vs. testing samples in the air above the site; direct measurement vs. modeling) to test emissions from shale gas drilling operations throughout the country. The EPA collects data from industry and reports this in a "Greenhouse Gas Inventory" Report each year. The agency has estimated very low leakage rates (0.16% to 1.4%), and stated in a draft 2015 report that total methane emissions from natural gas operations in the US declined by 9% between 1990 and 2013 (a sharp rise between 1990 and 2007 was followed by a 35% decline from 2007 to 2013).[44] Critics of these findings, such as Robert Howarth, whose findings in 2011 prompted a surge in research on this issue, have questioned EPA results and methods, arguing that there are much higher rates of methane leakage.[45] In 2013, a study conducted by University of Texas researchers suggested that the levels of methane leaking from all phases of natural gas development were about 10 percent below the estimate made by the EPA in 2011.[46] In one part of the drilling process, known as "completion," in which new wells are prepared for extracting gas, the study found that with proper controls in place and well-trained workers, 97 percent of the methane leaks could be captured and prevented from entering the atmosphere. This finding suggests that, "the oil and gas industry—when sufficiently motivated—can produce natural gas with modestly low emissions."[47] However, the study also found that at a number of existing wells already producing natural gas, the leakage was up to 67 percent higher than EPA estimates. Two additional studies conducted in 2014 disputed the notion that EPA estimates were accurate, suggesting instead that they were far too low as compared to testing done at a variety of operating sites. One study in particular measured concentrations of methane in the atmosphere above drilling sites in Pennsylvania and found that methane leaks were 100 to 1,000 times greater than the EPA has estimated to be the norm.[48]

The many studies conducted regarding the effects of hydraulic fracturing on water, chemical pollution and methane leaks have by no means ended debate, and it is not the aim of the brief summary of research presented here to assess their collective scientific merit. Additional investigations are likely to help determine more precisely any adverse impacts and the identification of effective ways to mitigate them. At the same time, such investigations will most certainly fuel additional disagreement and prolong the controversy. As these efforts both inform and shape the scientific understandings regarding America's energy profile, they also become part of the political and social dynamics surrounding the topic of energy security, generating a fair amount of acrimonious dispute in the process. As science and scientific research gets translated into politics and advocacy in the policy process, the elevation of seemingly narrow topics such as the determination of methane leaks from shale gas operations in Pennsylvania or the flow rate of chemicals underground exemplifies how unknown and esoteric subjects can quickly become critical components of the conversation about the nation's energy security. After all, an understanding of energy security necessarily rests upon a broad knowledge of the energy policy process, natural resource economics, international relations, and domestic politics, along with an ability to follow the ongoing research on the public health and environmental impacts of the ongoing energy transformation made possible by horizontal drilling and hydraulic fracturing.

The Policy Framework(s)

The array of policies that govern the oil and gas industry in the United States is a mixture of federal and state laws and regulations, requiring a variety of practices and permits in order to operate. At the same time, this policy environment also reflects the benefits of greatly increasing domestic supplies of oil and gas, and in general seeks to facilitate the development of these energy resources. Government entities can, and in some important ways, have chosen to largely "stay out of the way" and let the energy market develop. In such circumstances, market development has at times gotten ahead of scientific inquiry and public policy development, which is not surprising. In addition, in the same way that the oil and gas industry has fought to combat the belief that fracking is harmful, it has also sought to keep regulatory requirements at a minimum, and to preserve the relatively favorable regulatory regime that applies to it at the present time. In this context, what has become particularly notable, and rather controversial, is that there are practices and permits that one would expect to be required in oil and gas operations, but which are in fact not required. At the national level, federal environmental laws tend to be quite comprehensive. However, in several instances, oil and gas production are purposefully and specifically exempted from several important requirements that are designed to protect human health and prevent damage from being done to the environment.

The *Safe Drinking Water Act*, enacted into law in 1974, is the primary law that ensures the quality of the nation's drinking water and its principal sources. Among other things, the law regulates the injection of substances into the ground in order to protect groundwater from contamination. It is clearly the case that the process of extracting shale oil and gas involves sizable underground injections. However, this particular practice gets a regulatory exemption. The EPA completed a study in 2004, concluding that fracking posed little or no threat to drinking water, and that no further study of the issue was necessary.[49] The exemption was then written into law in the *Energy Policy Act of 2005*, which specifically defined "underground injection" to exclude "the underground injection of fluids or propping agents (other than diesel fuels) pursuant to hydraulic fracturing operations related to oil, gas, or geothermal production activities."[50] This provision became known as the "Halliburton loophole," so-named because then-Vice President Dick Cheney, a former CEO of Halliburton, pushed for its inclusion in the law (the name also stuck because it was Halliburton engineers who, decades earlier, had invented the process of hydraulic fracturing). The Vice President had also led an energy policy task force in 2001 (the National Energy Policy Development Group), whose report strongly supported additional oil and gas production, while recognizing the growing value of fracking and recommending that "we should reconsider any regulatory restrictions that do not take technological advances into account."[51] The task force report has been understood as the forerunner of the exemptions enacted into law in 2005. With this exemption, one important means of protecting drinking water aquifers from contamination from hydraulic fracturing operations has been taken out of federal hands and left to the individual states.

A second exemption that has generated considerable controversy is the exclusion of fracking wastes from the definition of "hazardous wastes" in the *Resource Conservation and Recovery Act of 1976* (RCRA). RCRA governs the safe disposal of solid waste and hazardous waste, but the law has exempted wastes from the exploration and production of oil and natural gas since 1980, well before fracking became a common practice. In 1988, an EPA regulatory report concluded that other existing federal and state laws (such as the *Safe Drinking Water Act* and the *Clean Water Act*) were sufficient to manage oil and gas wastes, and that adding a new requirement under RCRA would "cause a severe economic impact on the industry and on oil and gas production in the U.S."[52] A subsequent EPA report issued in 2002 further explained that these "special wastes" were considered low enough in toxicity to warrant an explicit exclusion under existing laws and derivative administrative regulations.[53] The result is that federal law does not regulate either the disposal of fracking wastes on land, or their injection into the ground.[54]

One of the most contentious elements of the policy framework governing unconventional oil and gas drilling is the federal exemption from disclosure requirements under the *Emergency Planning and Community Right to Know Act*. The statute requires that firms submit reports to the EPA detailing their use of toxic chemicals. However, the law does not require companies to disclose information about the chemicals they use for hydraulic fracturing, since this information can be considered a trade secret and can, as a consequence, be exempted from the reach of that otherwise broadly applied law. It can plausibly be argued that "creating a fracking fluid mixture that produces the desired results—that is, one that maximizes the ability of the well to produce gas for a sustained period—is indeed an extremely complicated task. The right mixture is both a valuable commodity to the owner, and the product of considerable intellectual effort and resources so it is natural for energy companies to try to protect that information."[55] At the same time, there is also a compelling public interest in knowing what types of chemicals are being used. As the National Resources Defense Council points out, the public, including state and local health and safety professionals, cannot fully understand the risks to air and water or accurately assess any impacts that may occur. NRDC argues that whether the issue is one of groundwater or surface water contamination, storage of flowback water, the release of known carcinogens into the air and/or water, or responding to blowouts or other accidents, the lack of disclosure leaves the public at risk, while depriving the public, regulators and scientists of the information they need to understand the possible effects of the chemicals and how they are used in domestic oil and gas production.[56]

Other such exemptions involve both the *Clean Water Act* and the *Clean Air Act*. The *Clean Water Act*, which covers the nation's rivers, streams, creeks, and wetlands (it does not focus on groundwater, which is dealt with in other laws), employs a permitting system, the National Pollution Discharge Elimination System, to minimize and control the discharge of harmful pollutants into the nation's surface waters. The law covers flowback waters, but it specifically exempts stormwater runoff (a potentially significant source of water pollution) resulting from all oil and gas construction and production activities.[57] The *Clean Air Act* regulates the emission of pollutants from both moving and stationary sources, and oil and gas production can produce toxic air pollution, including volatile organic compounds, hydrochloric acid, and hydrogen sulfide. It is usually the case that when "numerous small sources of air pollution, such as individual oil and gas wells and associated facilities, are under common control and in close proximity, they are treated as a 'major source' subject to stringent *Clean Air Act* technology requirements."[58] With regard to oil and gas production, their operations are specifically exempted from this aggregation requirement.

The various exemptions noted do not mean that oil and gas drilling is not heavily regulated, though it does seem that this industry is given a break in key areas that other industries do not enjoy. Nonetheless, where federal law has been absent, several US states have produced their own regulatory framework. It is important to note that the EPA carries out a large part of its responsibilities by working with state environmental agencies to implement and enforce federal environmental laws and regulations. So even though federal law leaves some areas unregulated, the EPA can and does work with states to carry out state policies that address some of the exemptions noted above. Still, a US Government Accountability Office review of the EPA's efforts to work with states in minimizing the impact on groundwater of underground injections issued in 2014 found that the agency is not taking sufficient action to carry out its responsibilities to oversee state regulatory programs and ensure their timely enforcement.[59]

Most state oil and gas regulatory frameworks have tended to focus on operational controls such as well design, drilling procedures, monitoring and oversight procedures, and the handling of materials and wastes. In recent years, the production of unconventional oil and gas has prompted some new state-level regulations involving disclosure of chemicals used, and the periodic testing of both surface and groundwater to assess impacts. In this regard, New York and Vermont have been the least amenable to hydraulic fracturing. New York, which includes a large part of the Marcellus Shale, banned the practice in 2014 after a study by the state health commission. This ban followed similar action taken by a number of the state's localities, which had been upheld by the New York State Supreme Court. Vermont went a step further than New York and passed legislation specifically banning the practice of fracking in the state. Most states, however, have enacted a variety of laws and regulations, resulting in a patchwork of different rules to be followed by energy producers. For example, Colorado requires disclosure of the chemicals used, public notice of proposed drilling activities, and the periodic testing of surface and groundwater both before and after drilling operations have taken place. On the other hand, California, until the passage of legislation in 2014, did not have any of these types of requirements. With respect to disclosure, Pennsylvania, Texas, Colorado, Wyoming, New Mexico, and Ohio have enacted varying requirements in these areas. On the other end of the continuum, in North Carolina legislation was passed in 2014 that lifted a ban on fracking that had been imposed in 2012 in order to develop a regulatory framework. The law makes it illegal for anyone, including health care providers and first responders to emergencies, to disclose information about the chemicals used in hydraulic fracturing.

Amidst a rapidly changing landscape, with battles being fought in numerous state legislatures, and a non-uniform set of policies developing across the country, there is a strong argument to be made for developing consistent, strict, enforceable national standards and practices to regulate this rapidly growing industry. In fact, George Mitchell himself, whose company made the breakthrough that enabled the contemporary oil and gas boom, before his death called for greater federal regulation of hydraulic fracturing. He pointed out that there are proper techniques that can and should be followed by every drilling company, stating "They should have very strict controls. The Department of Energy should do it."[60] This is a reasonable position to take. To the extent that people and the environment are harmed as a result of unconventional oil and gas production, public opposition will be likely to grow to the siting of new projects. Preventing such harm to begin with will assure the industry of greater social and political acceptance and the potential for building public support over the long term. Whether or not this occurs, the nature and level of federal and state regulation is expected to continue to change for some time into the future as experience with the technology grows and the science related to the documentation of outcomes improves. And of course, the science and the policy framework regarding the pursuit of unconventional oil and gas will continue to generate a great deal of political conflict, just as it has done in the past. The issues that arise around this subject are too big not to be of major political concern, as they are bound up in the larger debate over energy supplies, their costs and impacts, the role that the US government and the states can and should take in shaping the nation's energy profile, and the tradeoffs among economic, environmental and national security goals that are inherent in the systematic pursuit of energy security.

The Keystone Pipeline and the Politics of Oil

If there is one issue that has captured the essence of the debate over the role of the oil and gas boom in meeting or diminishing the goal of energy security, it is the Keystone XL pipeline. While the oil resources involved come from Canada, and are not acquired by using hydraulic fracturing technology, they are part of the larger expansion of oil production in North America, and as such the topic has been easily included in the ongoing conversation about the oil and gas industry transformation of recent years.

The politics of oil in the United States are acrimonious, to say the least. The current debate pits concerns over energy supply vs. those involving climate change, fossil fuels vs. renewables, and economic development, revolving around the question about whether the United States should further enable fossil fuel production or should instead focus its attention and investments on cleaner, alternative energy sources. The controversy over the Keystone XL pipeline embodies all these disparate elements, and has generated a level of intensity such that the two opposing sides have drawn a line in the sand to make this a signature issue about the country's energy choices. The debate has eclipsed the actual impact that the pipeline's construction or abandonment would ever have, but this is largely beside the point to the antagonists. As such, Keystone exemplifies and highlights all the competing elements of the energy security dilemma.

The proposed Keystone XL pipeline is an addition to the already existing 1,700 mile-long Keystone pipeline, which runs from Alberta, Canada to Cushing, Oklahoma. The expansion, shown in Figure 4.2, has been proposed by Trans-Canada Corporation, to add a new 875-mile pipeline from Alberta to Steele City, Nebraska.[61] An additional expansion, which runs from Oklahoma to Texas, became operational in January of 2014. The aim of the project is to send "heavy" oil shale produced in the Alberta tar sands, along with oil produced in the United States (including the Bakken Shale) to refineries in the United States.

As there are an estimated 170 billion barrels of oil in Alberta, the completion of the Keystone XL could effectively double the amount of oil imported from Canada, and could by itself send more than 800,000 barrels of oil per day to the United States, accounting for 9 percent of US petroleum imports.[62] At the same time, it could help address the problem of transporting oil coming out of North Dakota, which is currently transported by rail, trucks and barges that are far more prone to accidents than a pipeline. Because the project crosses an international border, the State Department has been responsible for the permitting and conducting the requisite environmental assessment, and ultimate responsibility has fallen on President Obama to grant or deny approval. However, due to the high stakes politics of the issue, the project has remained stalled for several years, with President Obama repeatedly delaying a final decision on approval since 2011.

The proposed pipeline seems able to carry with it not only oil, but also the hopes and fears of supporters and opponents alike. On the opposing side are environmental organizations and their allies, renewable energy proponents, many farmers and ranchers in Nebraska, Kansas and other states where the pipeline will be built, and a sizable number of Democratic members of Congress in both the House and the Senate. The State Department's environmental assessment was the original "hook" by which opponents have sought to stop or delay the project. Initial concerns included the passage of the pipeline through Nebraska's Sandhills region, which sits atop the Ogallala acquifer, one of the largest supplies of groundwater in North America.[63] The proposed route was revised to address this concern. Later objections have centered around impacts on wildlife and

FIGURE 4.2 The Keystone Pipeline
Source: US Department of State.

endangered species along the route, and the threat of spills on nearby waterways. The overarching issue, however, is that of climate change. As the argument was expressed in what has been called the *Keystone Principle*, "to avoid truly catastrophic climate change, the world needs to avoid new, long-term capital investments that are going to 'lock in' dangerous levels of carbon emissions" for years to come.[64] The world only has a limited capacity to absorb carbon, and the chance to preserve a stable climate would be lost forever if too much fossil fuel is extracted and consumed. Since the expansion of Keystone would do exactly this, encouraging and facilitating a large amount of an environmentally destructive form of energy, it should, goes the argument, be rejected. To that end, the environmental movement opposition to Keystone has come to stand out and partially eclipse other issues, while becoming a litmus test by which the President, along with other candidates running for office, are to be assessed on their commitment to environmentalism.

On the other side of the argument are the oil industry, business groups and their supporters, labor organizations (many construction and engineering jobs are expected to be created with the project), Congressional Republicans, and conservatives, along with a number of Democrats from states (especially "red" states) that want to see the pipeline built and North American oil production expanded. These supporters argue that the pipeline will help the American economy, adding new jobs and economic activity, while contributing to US energy security by augmenting oil supplies from a reliable source. In addition, it is argued that the production of Canadian oil will not substantially contribute to climate change— no single project like this could, considering the level of ongoing consumption worldwide. With respect to economic impacts, the analysis completed by the State Department found that the project will create roughly 42,000 jobs and add billions of dollars into the economy, though as critics point out, after construction of the pipeline, there would only be 50 permanent jobs created by the project. Nor, as stated in the State Department report, would it significantly increase greenhouse gas emissions in the United States.[65] Lastly, say pipeline proponents, even without the pipeline, oil from the Canadian tar sands would still be developed and sold, but most likely to other oil consumers such as China and India. This would be the worst outcome possible, forgoing enhanced energy security and economic growth while doing nothing to reduce carbon emissions.

President Obama has been stuck in the middle of this fight, sympathetic and beholden to the political left, which has consistently supported him, while also seeking to expand domestic energy production and help Democrats from oil and gas producing regions of the country. The President has repeatedly put off a final decision, calling at various times for revisions to the plan, further study, the passage of the 2012 elections, and final rulings on litigation in Nebraska involving the pipeline route. In 2015 Congress passed legislation approving construction of the pipeline, but President Obama vetoed the bill, and his veto was not overturned. The President has stated that ultimately he will oppose the project if it significantly increases carbon pollution, but since both sides can make a plausible argument about what "significant" actually means, the President has left himself options while allowing the debate to rage on. Regardless of President Obama's final decision (which is not yet known at this writing), the issue is not likely to disappear and will be taken up by his successor.

A Move away from Coal?

Any set of policies designed to transform the energy landscape and enhance energy security in the United States has to address the outsized role that coal plays in the nation's electric power sector. Coal is a domestic energy source that exists in great quantities. There are estimated to be 270 billion tons of recoverable reserves in the United States, more than 25% of the total known coal reserves in the entire world.[66] This is enough coal to last for more than two centuries at current rates of use (though such projections do not take into account exports or the potential for growing usage if carbon capture and storage technology becomes widely available). In 2013, the United States mined 1.02 billion tons of coal, burning 874 million tons of it (86%) in domestic power plants to produce 1.6 trillion kilowatt-hours of electricity, or 39% of total power generation. Exports accounted for 107 million tons.[67] The high point of coal mining in the United States came in 2008, when 1.2 billion tons were mined. With respect to generation, the years 2000–2008 saw a high point of roughly 2 trillion kilowatt-hours of electricity generated from coal, ranging from 48 to 50 percent of the country's total.[68]

Because coal is so abundant in the United States, it has long been an important energy source and will continue to be for a long time into the future. It offers numerous energy security benefits, it is abundant and affordable, the supply chain is reliable, and its acquisition does not jeopardize US national security in the way that oil does. As part of a mix including natural gas and nuclear energy, coal can contribute to maintaining a diverse supply of resources to power growing electricity demands for a very long time to come. The three fuels provided 85% of all US electricity in 2014; by contrast, wind and solar provided less than 5%.[69]

Despite these noteworthy advantages, the amount of coal used in the American electricity sector has declined in recent years due to a number of policy-related and financial challenges, resulting principally from the fact that coal is the dirtiest of all fossil fuels. Mining and burning coal adversely impacts the environment and public health in well-documented ways. Even when mined, transported, and burned as intended, in compliance with established legal and regulatory requirements, coal usage contributes to air and water pollution, poorer health of miners and populations in nearby communities, and global climate change. Industrial accidents, of which there are numerous examples every year involving incidents such as mine accidents, chemical spills, and impoundment dam breaches, make the adverse public health and environmental impacts even worse.

Compounding the environmental and health concerns, there are economic forces at work affecting the continued use of coal. There is the growing availability and relatively low price of natural gas, driven by hydraulic fracturing of shale gas. This energy industry transformation has prompted developers and utilities to increasingly prefer building natural gas power plants over coal plants. At the same time, the American economy has not been in top form since the 2008 financial crisis, and this has kept energy use from growing as rapidly as it had in earlier years.

Amidst all these growing concerns and developments there have been a number of policy measures—adopted and proposed—designed to diminish the environmental impacts of coal, and to reduce its use overall. These substantial changes have all led to a reduction in coal use and a looming uncertainty about its future in the US energy mix.

Regulating Coal

There are a multitude of policy measures primarily designed to mitigate the harm that comes from coal mining and combustion. The coal industry is one of the most heavily regulated in the United States. Most of this regulation has been developed since the 1970s, when two signal developments—the growing environmental movement and the oil embargo of 1973—prompted efforts in the United States to both increase the use of coal in electricity production, and to mitigate the environmental impacts of using coal for this purpose.

Surface mining, which provides 70% of the country's coal, is governed by the *Surface Mining Control and Reclamation Act of 1977* (SMCRA). It is implemented and enforced by both the states and the Office of Surface Mining Reclamation and Enforcement in the US Department of Interior. Underground mining is covered under the *Mine Safety and Health Act of 1977*, and is regulated by the Mine Safety and Health Administration in the Department of Labor. Mining operations also have to comply with provisions of the *Clean Air Act*, the *Clean Water Act*, the *National Environmental Policy Act*, the *Toxic Substances Control Act*, the *Safe Drinking Water Act*, the *Emergency Planning and Community Right to Know Act*, and the *Environmental Response, Compensation and Liability Act*, to name only some of the most relevant statutes, all of which authorize a variety of regulations applicable to coal mining.

The regulation of power plants is largely the work of the EPA, whose enforcement of the *Clean Air Act* provides a key tool in managing air pollution in the nation. Under the *Clean Air Act*, the EPA has implemented a number of regulations placing limits on the levels of sulfur oxides (SOx), nitrogen oxides (NOx) and fine particle matter that can be emitted, requiring scrubbers to trap these pollutants along with other known contaminants. At the same time, this regulatory regime is being strengthened and extended. The Mercury and Air Toxics Standards (MATS) are some of the most recent additions to the regulatory framework governing coal power plants. Initiated in 2011 and under ongoing revision, the MATS were adopted after several years of development. Up until that time, there had been no federal standards requiring power plants to limit their emissions of mercury, lead, arsenic, nickel, cobalt, and other toxic gases.[70] The regulation faced a legal challenge and in June 2015 the Supreme Court ruled that the EPA did not properly consider the costs of implementing the measure as it developed the rule. The ruling did not invalidate MATS altogether. Instead the issue was sent back to the DC Circuit Court of Appeals to determine whether the EPA could revise the rule to address the Supreme Court's ruling.

Another measure, the Cross-State Air Pollution rule, finalized in 2011, is designed to cut emissions of SOx and NOx that carry far beyond generation plants. States are required to be "good neighbors," controlling emissions so that they do not travel downwind and diminish air quality in a neighboring state. This cross-state airborne pollution can prevent other cities and regions from maintaining EPA-enforced National Ambient Air Quality Standards, and cause adverse public health impacts.[71] While the Cross-State Air Pollution rule was also challenged in court, and had been set aside by the DC Circuit Court in 2012, the Supreme Court in 2014 overturned the Circuit Court's decision, ruling that the EPA does in fact have the authority to promulgate and enforce these regulations.

Another one of the key regulatory tools regarding power plants has been the New Source Review (NSR) permitting program. Established in the 1977 *Clean Air Act Amendments*, NSR requires pre-construction approval for new power plants, and for major modifications to existing plants, to ensure that air pollution standards are met. In effect, this means that the EPA, along with state and local pollution control agencies, can do things such as require the use of certain scrubbers or other equipment (termed "Best Available Control Technology"), specify the height of smokestacks, and decide what types of pollutants can be emitted at what levels.[72] A particularly contentious issue surrounding the NSR is defining what types of modifications trigger a new permitting process and compliance with stricter standards. A provision in the *Clean Air Act* "grandfathered" old power plants that lacked certain pollution controls, making them subject to NSR standards only at the time that they made significant upgrades, expansions, and/or major modifications. As a result, many coal-fired plants in the United States do not use the Best Available Control Technologies. For example, in 2010 coal plants that did not use flue gas desulfurization scrubbers produced 42% of the electricity generated from coal, but accounted for 73% of coal's SOx emissions.[73] However, because clear and consistent interpretation of what triggers stricter compliance has been elusive, the EPA has issued differing interpretations and a variety of rules (Democratic administrations have pushed for stricter standards, while Republicans have tended to side with the utilities, who want to avoid costly pollution controls), and rival lawyers for power utilities and environmental organizations have gone through countless lawsuits trying to either preserve exemptions or force stricter NSR regulations and compliance. The 2014 Supreme Court ruling on the Cross-State Air Pollution rule is expected to give EPA the authority to require new pollution control technology in some plants that had previously been grandfathered under the NSR provision.

Within this legal and regulatory framework, which is clearly extensive, there continues to be an effort to further address the environmental impacts of the use of coal, and this is part of the ongoing attempt to achieve a transformation in the American energy profile. With regard to coal-fired power plants, one area in particular that has not been historically addressed is the emission of carbon dioxide. To that end, policy initiatives have been undertaken to combat climate change by means of modifying coal and coal plants. The two primary ways this is being pursued are via the regulation of carbon emissions through the EPA under the *Clean Air Act*, and in supporting the development of "clean coal" technologies, otherwise known as carbon capture and storage (CCS).

The EPA and Carbon Dioxide

As the subject of climate change and carbon dioxide emissions came to acquire greater saliency, the question arose regarding the authority of the EPA to regulate such emissions as an air pollutant. In the early 2000s Massachusetts and several other states petitioned the EPA to do this. EPA declined their request, citing the fact that Congress had not granted the agency authority to regulate greenhouse gases, and that doing so in the ways being proposed would have significant detrimental economic, practical and societal impacts, while simultaneously adversely affecting other policies used to address the problem. The issue went to court, and in 2007 the US Supreme Court ruled in *Massachusetts v. Environmental Protection Agency* that the EPA did have the authority to regulate carbon dioxide. In effect, the Court ruled that greenhouse gases meet the definition of air pollutants under the *Clean Air Act*. Moreover, the Court ruled that the EPA was obligated to regulate CO_2 if its emission could be reasonably anticipated to endanger public health or welfare. Based on this major ruling, and following the election of President Obama, the EPA began the effort to regulate carbon dioxide emissions from both automobiles (as part of revised CAFE standards issued in 2011), and of power plants, both new plants and existing ones.

Starting with new power plants, the EPA issued a proposed rule in April 2012 to establish a "New Source Performance Standard" for carbon that would place a limit of 1,000 pounds of CO_2 emissions per megawatt-hour of electricity generated.[74] This standard would apply to both coal-powered and natural gas–powered plants. A controversy erupted immediately because this standard is feasible only in natural gas plants. The carbon capture and storage technology that would be required for coal plants to meet the standard is so new and so expensive, that it would effectively mean the end to new coal plant construction, at least until CCS technology has been proven to be scalable and cost-effective. At this writing, no one can say for certain if or when CCS will meet such milestones, even though much work is being done to get beyond the proof of concept stage of development. The resulting opposition from the coal and utility industries to this proposal was quick and forceful, and it prompted the EPA to try again to meet its court-imposed obligation to regulate greenhouse gases under the *Clean Air Act*. In September 2013 a revised proposal was made, with a new standard set for coal plants. While natural gas plants would have to meet the cap of 1,000 lbs per megawatt-hour, coal plants would have two options to be ruled in compliance. Neither option would eliminate the need to use CCS technology once it is available, but they would grant additional time to become compliant with standards set. Under one option, coal plants would have to start using CCS soon after startup to achieve a 12-month average emission rate of 1,100 lbs of CO_2 per megawatt-hour. A second option would permit coal plants to begin using CCS much later to achieve a seven-year average emission rate of between 1,000 and 1,050 lbs of CO_2 per megawatt-hour.[75] The final rule went through an extensive review process and went into force in 2015.

While the agency's rules were being developed for new power plant carbon emissions, the EPA began a concurrent effort to create a standard for existing power plants. In June 2014, it released a proposed rule, called the Clean Power Plan, to cut CO_2 emissions from existing power plants nationally by 30% from 2005 levels by 2030.[76] The rule leaves the exact mechanism for achieving compliance up to each state, which can meet standards through a combination of "building blocks"—such as documented energy efficiency enhancements, demand management, increased reliance on low-carbon or zero-carbon fuels, and improvement in power plant operations. There is a fair degree of complexity in the rule (the final rule published online by EPA in August 2015 was 1560 pages in length, though this included a full accounting of the rulemaking process and interpretations of the statutory authority under which the rule was developed), with each state's target being set as a carbon intensity rate as opposed to a total emissions level as specified in the New Source Performance Review. The rule does not apply to each power plant, and each state's target is different based on their starting point with regard to factors such as their overall fuel mix, the level of use of renewables, and policies already in place (the standard is termed an "Adjusted MWh-Weighted-Average Pounds of CO_2 per Net MWh from all Affected Fossil Fuel-Fired Electricity Generating Units"). The core idea of the new regulatory framework for power plants is to aim for what states can realistically achieve using existing technologies and best practices at a reasonable cost. For example, Arizona has been making a concerted effort to promote solar power, while the closure of Oregon's only coal-fired plant in the city of Boardman, scheduled for 2020, should bring the state into full compliance with the rule. On the other hand, states such as Kentucky and West Virginia have more modest targets because their power generation systems have less potential for major reductions. Kentucky's goal by 2030 is 1,286 pounds of CO_2 per megawatt-hour, while Oregon's is 871 pounds.[77] It should also be noted that due to the increasing use of natural gas, the diminished use of coal, the promotion of renewables at the state and federal levels, and the economic downturn, there have already been a number of changes that have led to carbon intensity reductions in several states below 2005 levels. In other words, the proposed rule counts some of the progress already made as part of its total carbon reduction goals. If states do not offer their own compliance plans, the EPA will impose a federal plan. Like the standard for new power plants, the Clean Power Plan was finalized in 2015.

Not unsurprisingly, the EPA's move has been hailed by its supporters as one of the most significant actions the US government has ever taken to address global warming, and decried by opponents as a major threat to the US economy. Former Vice President Al Gore said that the rule is "the most important step taken to combat the climate crisis in our country's history," while the American Coalition for Clean Coal Electricity said that "if these rules are allowed to go into effect, the (Obama) Administration is for all intents and purposes creating America's next energy crisis."[78] The EPA's estimate is that this rule will increase the cost of electricity by about 3% on average around the country by 2030. However, it expects that gains in efficiency will result in less electricity usage and an overall reduction in homeowners' utility bills by 9%.[79] Added to these savings are the expected reductions in health care costs associated with cutting air and water pollution from using coal fired electricity on a large scale.

The regulatory regime that is evolving—incorporating new pollution control technologies, setting standards for mercury and other toxins, and placing caps on carbon emissions—can be reasonably understood as an effort to reduce the environmental and health impacts of using coal as a primary source of energy. These effects represent the one major downside of using coal, which otherwise is abundant, affordable and reliable. Neutralizing these effects, or at least reducing them to acceptable levels, on par with other energy resources, especially those with the least environmental and health impacts, is a worthy goal indeed. Success in this realm could make coal an ideal energy source for the United States vis-à-vis energy security. However, because the road to achieving these successes, which are uncertain at best, can be costly to everyone in the supply chain (the mining industry, power plant owners, utilities, and consumers), there is a certain point where the EPA's effort to mitigate health and environmental effects can look more like an effort to end the use of coal altogether. In this regard, the EPA and the Obama Administration have been accused often of waging a "war on coal" by their critics, particularly in coal-producing states, as a result of the stricter regulatory environment and the proposed carbon emission limits. (A Google search of "Obama War on Coal" yielded more than 15 million results.) This view is strengthened by the Administration's own words, which state in *The President's Climate Action Plan* that, "[g]oing forward, we will promote fuel-switching from coal to gas for electricity production."[80]

The argument made by critics of the new rules is that limiting coal use will depress local economies in coal mining states such as West Virginia, Kentucky, Wyoming and Pennsylvania; cause electricity prices to rise nationwide; and do little to improve the national and global environment. While the first of these claims seems beyond dispute, the second and third are questionable. Inexpensive natural gas, along with the falling costs of wind and solar power, are strongly driving growth in the use of these energy sources. And the environmental impacts of coal mining and combustion are so widespread and well-understood that reducing coal use seems certain to have an overall beneficial effect.

To combat the changing policy environment, coal's supporters, both in industry and in government, are strongly defending their interests and pursuing a goal of neutralizing or reversing these policy changes. They are engaging in public relations campaigns and in political efforts in several states to reduce or eliminate incentives and mandates promoting wind and solar power. For example, legislative proposals in Kansas, North Carolina and Arizona would revise net metering laws, impose fees on owners of solar PV systems, and/or repeal renewable portfolio standards. There are also legal challenges to the New Source Performance Standard and the Clean Power Plan, including a lawsuit by several states challenging the Clean Power Plan. In one case, *Utility Air Regulatory Group v. Environmental Protection Agency*, the plaintiffs claim that the EPA has overstepped its authority in carrying out the Supreme Court's 2007 ruling.[81] (This legal approach is potentially a smart one, saying that the problem is not the law itself—though one expects the plaintiffs believe it is—but rather the way it's being enforced.) Utilities and mining interests, along with conservative organizations such as the American Legislative Exchange Council (a non-profit organization consisting of conservative state legislators from around the country), have supported these efforts, which are grounded primarily in financial interest, but also in an ideology—embodied in the Tea Party and much of the Republican Party—that seeks to reduce the size and scope of government action in general.

Senator Mitch McConnell of Kentucky has pursued an additional approach to oppose the growing regulatory measures involving coal. He sent a letter in March 2015 to the National Governors Association stating that the EPA is "attempting to shut down America's coal-fired power plants," and he encouraged states to refuse to cooperate with the EPA in developing strategies to comply with the Clean Power Plan. Stating that he fully expects the plan to be invalidated in court, McConnell argued that the EPA is imposing a short deadline in the hope that "states will commit to these plans before courts can decide on the legality of the CPP." Therefore, "[g]iven the dubious legal rationale behind the EPA's demands, rather than submitting plans now, states should allow the courts to rule on the merits of the CPP."[82] In spite of Senator McConnell's efforts, states are unlikely to want to ignore the regulation, and the EPA has reported that it is working closely with regulators in many states, including those suing the EPA, to develop their compliance plans.[83] The political battle surrounding this issue is intense because the stakes are very high, and it is unclear whose interpretation of the issue—those who speak of "a war on coal" conducted through "job-killing regulations" or those who talk about cutting "carbon pollution" and "protecting public health and the environment"—will ultimately prevail. For the time being, however, the move to create a stronger regulatory regime for coal is making major inroads.

Should the Clean Power Plan for existing plants come fully into force, it is expected that a real reduction in carbon emissions could be achieved, along with a reduction in all the other pollutants emitted from coal-fired energy. With respect to the New Source Performance Standard, however, the practical effect of the policy changes may turn out to be rather minimal. Due to increased production and low prices of natural gas in the United States, there are very few new coal-fired plants either being planned or under construction. The US Energy

Information Administration reports that almost all proposed new fossil fuel generation capacity for the next several years is expected to be gas-fired.[84] In fact, there are several planned closures of coal plants nationwide in the states of Oregon, Georgia, Kentucky, Michigan, South Carolina, Massachusetts and possibly others.[85] So even without the EPA's efforts to regulate carbon emissions from power plants, it is unlikely that many new coal plants would be built in the decade after the rule was promulgated. Nonetheless, it can credibly be argued that the changing and uncertain regulatory actions (along with growing public opposition to coal use in parts of the country, which is both a cause and a consequence of policy change), will continue to play a growing role in prompting planned coal plant retirements and the move toward greater use of natural gas and renewables for electric power production.

As the Obama Administration comes to a close, the policy environment is one in which a concerted effort is underway to reduce coal use because it is harmful to public health and the environment, while also placing additional regulatory controls on its use to reduce the externalized costs, in recognition of the fact that coal will remain an important energy resource for a long time. At the same time, another government effort is being pursued on a parallel track. It involves supporting and investing in the development of carbon capture and storage technology, to see if it can become a technically and economically feasible option for the electric power sector in the future.

Carbon Capture and Storage, a.k.a. "Clean Coal"

The promise and potential of carbon capture and storage (CCS) technology is vast, offering the possibility of removing up to 90% of the carbon from fossil fuels used at both new and existing power plants. To the extent that cost-effective, scalable technologies to remove carbon from power plant emissions can be developed, along with improved transportation and storage capabilities, it will become possible to use coal and natural gas as largely carbon-free energy sources. Moreover, CCS is being pursued as a means to capture CO_2 emissions from other industrial processes such as producing steel, cement, chemicals and fuels. To that end, US policy is aimed toward partnering with industry to engage in research and development, and to help finance demonstration projects to test and prove this new, potentially promising technology and spur market development for it.

CCS is a process that is already used in several industries, such as producing ethanol, processing natural gas and hydrogen, and producing fertilizer. The captured CO_2 is employed in various ways, with one of the most well-known being enhanced oil recovery, whereby CO_2 is injected deep underground to push oil toward the wellhead. There are over 100 CCS projects in the US that use CO_2 in oil production, and this existing know-how and infrastructure is an advantage that CCS advocates point out in arguing the technology's feasibility.[86] However, while these industrial processes lend themselves well to separating CO_2 from other materials, the ability to do this in power plants is not well-developed. Nor has the ability to transport and store what would amount to very large volumes of carbon dioxide been demonstrated. The scale of operations for the electric power sector is much larger, and the process of separating out the CO_2 can be far more complex and expensive than current carbon capture technologies can achieve. More than twenty projects to develop power plants with CCS have been identified, proposed, studied and planned worldwide, but to date, only one in the world has begun operation, the Boundary Dam project in the province of Saskatchewan, Canada. One other project is currently under construction, the Kemper County project in the state of Mississippi, which is expected to become operational in 2016. All others are in various stages of planning.[87]

There are three major types of carbon capture technologies under active development. Pre-combustion carbon capture involves converting the coal into a gaseous mix of hydrogen and carbon dioxide. The hydrogen is then burned to produce electricity, while the CO_2 is collected for storage or industrial uses. Post-combustion capture entails separating the CO_2 from other exhaust gases after the coal is burned, using a solvent to absorb the carbon dioxide. The oxyfuel process is similar to post-combustion capture, except that instead of burning the coal in an air-filled chamber the system uses pure oxygen, which simplifies the collection of CO_2. Transporting the CO_2 will require a network of pipelines, due to the large volume that will be produced, though shipping some of it in tankers (via land or sea) may be viable options too. Industry advocates suggest that networks of hubs and storage clusters would be most likely to develop, linking power plants to storage sites. There are about 4,000 miles of pipelines in the United States used in transporting CO_2, so there is some infrastructure in place upon which to expand.[88] Finally, storage is likely to involve primarily pumping the CO_2 to deep, underground geological formations. However, there may be commercial uses that would turn the CO_2 into a marketable asset instead of a waste product. Carbon dioxide is currently used in enhanced oil recovery, and it is hoped/expected that the expanded volume of CO_2 can contribute further to domestic oil production. Other possibilities involve using the material in other industrial production processes. (Startup companies are experimenting with a variety of ideas, such as turning the CO_2 into bricks, or using it as feedstock to grow algae for biofuels. The key task, of course, is to keep the CO_2 from entering the atmosphere as a greenhouse gas.) Each of these steps in the process of capturing and storing and/or making productive use of CO_2 still requires a great deal of research, investment, testing and proof of concept demonstration to become technically viable at a scale needed to accommodate the electric power sector, and to determine whether they are financially feasible.

Due to the potentially sizable benefits of CCS, the US government has committed itself to the technology's development. CCS complements the energy security and climate change objectives of the nation, and it is likewise essential to the regulatory structure that the EPA is pursuing. Without it, the carbon dioxide New Source Performance Standards will effectively become a permanent ban on new coal-fired plants.

Since 2005, Congress has appropriated more than $6 billion for clean coal programs, committing a combined $2.2 billion in the *Energy Policy Act of 2005* and the *Energy Independence and Security Act of 2007*. It committed an additional $3.4 billion in the *American Recovery and Reinvestment Act of 2009* (ARRA), which represented the government's single largest investment in CCS, and represented a major push to greatly accelerate the R&D process. Since this time through 2013, Congress appropriated additional funds totaling over $1 billion.[89] Also after the passage of the ARRA, the Department of Energy's National Energy Technology Laboratory (DOE/NETL) developed a CCS "Roadmap," featuring a vision of "having an advanced CCS technology portfolio ready by 2020 for large-scale CCS demonstration that provides for the safe, cost-effective carbon management that will meet our Nation's goals for reducing GHG emissions."[90] Achieving this goal is an ambitious undertaking, as "DOE/NETL estimates that using today's commercially available CCS technologies would add around 80 percent to the cost of electricity for a new [pulverized coal] plant, and around 35 percent to the cost of electricity for a new advanced gasification-based plant." Therefore, "[t]he CCS RD&D effort is aggressively pursuing developments to reduce these costs to a less than 30 percent increase in the cost of electricity for [pulverized coal] power plants and a less than 10 percent increase in the cost of electricity for new gasification-based power plants."[91] What this has meant in a practical sense is a focus since 2009 on learning as much as possible as quickly as possible from large-scale demonstration projects.

The Kemper County project, which is owned and operated by the Mississippi Power Company, is the first of several planned large-scale projects. It will employ a pre-combustion, integrated gasification combined cycle (IGCC) technology. The plant will have 583 megawatts of generation capacity, and is expected to capture 65% of its CO_2 emissions. This will make the Kemper plant equivalent in emissions to a new natural gas-fired combined cycle plant, and will allow it to meet the EPA's CO_2 standard. The US Department of Energy provided $270 million, or roughly 10% of the planned $2.4 billion total cost. However, in April of 2013 the company announced that the actual project costs would be closer to $3.4 billion, far higher than the original estimate.[92] While such cost overruns are typical in the early attempts to demonstrate new technologies, the project is likely to raise further questions over the actual cost/value of the environmental benefits of CCS technology. Still, the experience with the Kemper project seems unlikely to cause DOE to significantly revise its long-term plan for CCS.

DOE's *Roadmap* called for not only additional power plant demonstration projects, but also for research, development and demonstration of industrial carbon capture (steel and cement production, for example, produce large quantities of CO_2); the reuse of CO_2 in producing fuels, fertilizer, cement, plastics and other goods; geologic storage; and a project called FutureGen.[93] FutureGen was initially proposed in 2003 as a $1 billion effort to create a completely new state-of-the-art CCS facility. The concept has been revised several times since then, and is still quite far from development. The plan as of 2015 is to retrofit an oil-fired plant in Illinois with an oxyfuel system. All of these R&D activities have led DOE to commit over $3.5 billion for 34 different CCS projects.[94]

The ultimate outcome of this joint effort by government and industry remains uncertain. This, of course, is the point of government support for industry in the early stages of technology development. No one knows if CCS will prove technically and financially feasible at levels to make a large impact. Still, the big investment by government and industry faces obstacles. CCS is expected to make power generation less efficient, as the technology requires energy to run, leaving less electricity to sell to customers. In addition, the cost of CCS may limit its attractiveness. Several studies have estimated a range of $30 to $85 per ton to capture CO_2, which would raise the price of electricity from 2 to 6 cents per kWh.[95] Moreover, ongoing technology and market development for renewables, natural gas, and nuclear power (or other energy sources) may change the relative value of CCS technology. As one analyst and advocate for renewables argued, "I am doubtful that CCS will ever pay. The cost curves for renewable power suggest that solar and wind will undercut the cost of CCS on new or retrofitted gas and coal plants before 2020, when CCS proponents hope that it will become economically viable."[96] Lastly, the heavy focus on capture technologies may result in a situation in which the ability to capture CO_2 outpaces the capability to store it all. Storing 20% of global carbon dioxide emissions would require pumping down twice the volume that the world pumps up in oil every year.[97] Until such a time that these variables and their costs can be more clearly determined, the Department of Energy will continue to make a big bet on the potential of CCS. As the Congressional Research Service notes, the coal program represented 70% of all fossil energy R&D appropriations from 2012 through 2014. In other words, "CCS has come to dominate coal R&D at DOE."[98]

Principles, Profits and Other Competing Interests

The changes occurring with respect to fossil fuels in the United States suggest that while there may be a long-term effort to step away from fossil fuels, especially coal, or at least to diminish their harmful impacts, this will not weaken the immediate efforts to fully exploit the opportunities that have become available to secure ever greater quantities of domestically-produced oil and natural gas.

As natural gas remains abundant and affordable as a power generation fuel in the United States, as policy regarding coal embodies stricter requirements, as long as carbon capture and storage technology remains in the demonstration phase, the prospects for a "coal renaissance" in the United States are still rather slim. However, overseas markets may pose few such limitations. Like the market for

American cigarette manufacturers, as public health restrictions grew to major proportions, the domestic market contracted but the export market grew in direct proportion. The US coal industry is likewise likely to see its growth come from overseas as domestic use is systematically displaced by natural gas, and perhaps wind power. The United States, as a matter of policy and practice, may be seeking to diminish its contribution to polluting the planet—in and of itself a noble principle—but what amounts in effect to exporting air pollution and greenhouse gas emissions will not address the global challenges of slowing climate change or improving air quality and reducing atmospheric pollution. Such actions will probably make these problems worse, using fossil fuels to ship coal across the oceans, often to places where pollution control standards and technologies are less prevalent than or nowhere near as strict as they are in the United States. But at the same time, cheap coal can mean abundant, affordable electricity for people who may not have access to it, providing opportunities for economic progress and a better quality of life.

With respect to the circumstances surrounding domestically-produced oil and natural gas, the economic benefits derived (and expected) are significant, as the United States comes to supply a larger share of the world's fossil fuels demand. Billions of dollars are at stake, along with (by some estimates) hundreds of thousands of jobs over the next several decades. At the same time, the environmental costs—on land, air and water—of acquiring shale gas and "tight" oil via fracking are becoming increasingly evident. These concerns are spurring research, clearer knowledge and deeper understanding of the impacts, proposals to limit the harm, and hope that policies and practices will soon change. However, at least for the immediate present, the short-term benefits, and perhaps longer-term benefits, are too important and too tempting to pass up for oil and gas companies, for policymakers at all levels of government, and for consumers alike.

While these developments are upending the country's fossil fuel energy profile, there is another energy boom occurring simultaneously, in the realm of renewable energy, particularly with regard to wind, solar and biofuels. It is the consideration of the public policy and domestic and global market changes surrounding these energy sources, and the transformation they are driving in the United States, that we turn to in the next chapter.

Notes

1 Michael Levi, *The Power Surge: Energy, Opportunity, and the Battle for America's Future*, New York: Oxford University Press, 2013.
2 Levi, p. 24.
3 For example, the US Energy Information Administration's *Annual Energy Outlook* for 2002 and 2003 both state that net imports of natural gas were expected to increase through 2020.
4 US Energy Information Administration (EIA), "Henry Hub Natural Gas Spot Price," www.eia.gov/dnav/ng/hist/rngwhhdW.htm (accessed April 25, 2015).

5 The terms "hydraulic fracturing" and "fracking" tend to be used in two different ways. On the one hand, they refer only to one part of the production process, whereby water, chemicals and sand are injected underground and then explosive charges open up shale rock formations. However, they can be used more broadly to refer to the entire process of shale oil and gas production. We are employing the broader usage here.

6 "Father of the Fracking Boom Dies," *Forbes*, June 27, 2013.

7 Detailed information on the location and output of shale oil and gas wells in the US can be found at http://shalebubble.org/the-map/ (accessed April 25, 2015).

8 EIA, *Annual Energy Outlook 2014*.

9 North Dakota Department of Mineral Resources, "Historical Monthly Oil Production Statistics," www.dmr.nd.gov/oilgas/stats/historicaloilprodstats.pdf (accessed October 10, 2015).

10 EIA, "Texas Field Production of Crude Oil," www.eia.gov/dnav/pet/hist/LeafHandler.ashx?n=PET&s=MCRFPTX2&f=M (accessed June 12, 2014); and David Blackmon, "Texas Continues to Lead the Shale Oil and Gas Revolution," *Forbes*, October 8, 2013.

11 International Energy Agency, *World Energy Outlook 2014*; EIA, "U.S. Expected to be Largest Producer of Petroleum and Natural Gas Hydrocarbons in 2013," *Today in Energy*, October 4, 2013, www.eia.gov/todayinenergy/detail.cfm?id=13251 (accessed June 12, 2014).

12 EIA, Petroleum Overview, *Monthly Energy Review*, April 2015, p. 45.

13 *Bloomberg News*, "Unforeseen US Oil Boom Upends Markets as Drilling Spreads," January 8, 2014, www.bloomberg.com/news/2014-01-08/unforseen-u-s-oil-boom-upends-world-markets-as-drilling-spreads.html (accessed June 12, 2014).

14 EIA, "U.S. Shale Production," www.eia.gov/dnav/ng/hist/res_epg0_r5302_nus_bcfa.htm; and Natural Gas Gross Withdrawals, www.eia.gov/dnav/ng/hist/n9010us2A.htm (accessed April 25, 2015).

15 US Energy Information Administration, "Natural Gas Gross Withdrawals and Production," www.eia.gov/dnav/ng/ng_prod_sum_dcu_NUS_a.htm (accessed April 25, 2015).

16 US Energy Information Administration, *Annual Energy Outlook 2015*.

17 EIA, "Henry Hub Natural Gas Spot Price," www.eia.gov/dnav/ng/hist/rngwhhdW.htm (accessed September 25, 2015).

18 A close friend of one of the authors managed the operation of directional drilling rigs throughout the United States for several years. At the time when oil prices reached a peak of $145/bbl in 2008, this person earned as much as $2,500 per day. The drop in prices to $50/bbl in early 2015 meant a sharp drop in drilling new wells, and the subsequent layoff of the author's friend and most of his colleagues who worked in these drilling operations.

19 Levi, pp. 26–27.

20 IHS, *America's New Energy Future: The Unconventional Oil and Gas Revolution and the US Economy*, 3, September 2013, pp. 47–55.

21 IHS, p. 41.

22 PricewaterhouseCoopers and the American Petroleum Institute, *Economic Impacts of the Oil and Gas Industry in 2011*, July 2013, pp. 6–7.

23 The United States Conference of Mayors, *U.S. Metro Economies: Impact of the Manufacturing Renaissance from Energy Intensive Sectors*, March 2014; IHS, *America's New Energy Future: The Unconventional Oil and Gas Revolution and the US Economy*, Vols. 1–3; PricewaterhouseCoopers, *Shale Gas: A Renaissance in US Manufacturing?* December 2011.

24 The United States Conference of Mayors, p. 1.

25 Blake Ellis, "Double Your Salary in the Middle of Nowhere, North Dakota," *CNN Money*, October 20, 2011, http://money.cnn.com/2011/09/28/pf/north_dakota_jobs/ (accessed May 1, 2015).

26 See, for example, J.S. Gilmore, "Boom Towns May Hinder Energy Resource Development," *Science*, 191, 1976, pp. 535–540; C.F. Cortese, "The Social Impacts of

Energy Development in the West: An Introduction" *The Social Science Journal*, 16, 1979, p. 2; and Jeffrey Jacquet, *Energy Boomtowns and Natural Gas*, NERCRD Rural Development Paper No. 43, 2009.

27 This is often referred to as "Stein's Law."

28 Mara Van Ells, "A North Dakota Oil Boom Goes Bust," *The Atlantic*, June 27, 2015, www. theatlantic.com/business/archive/2015/06/north-dakota-oil-boom-bust/396620/ (accessed September 24, 2015); "In Some Texas Oil Towns, This 'Downturn' Feels More Like a 'Bust,'" *StateImpact/NPR*, August 17, 2015, https://stateimpact.npr.org/texas/2015/08/17/in-some-texas-oil-towns-this-downturn-feels-more-like-a-bust/ (accessed September 24, 2015).

29 Nikos Tsafos, "The Missing Shale Miracle," *Foreign Affairs*, March 23, 2014.

30 Chris Mooney, "The Truth about Fracking," *Scientific American*, November 2011, pp. 80–85.

31 Hilary Boudet, Christopher Clarke, Dylan Bugden, et al., "'Fracking' Controversy and Communication: Using national survey data to understand public perceptions of hydraulic fracturing," *Energy Policy*, February 2014; Ben Schiller, "'Fracking' Comes to Europe, Sparking Rising Controversy," *Yale Environment 360*, April 2011.

32 Adam Reed, David Bernell, Jackson Cassady, et al., "Hydraulic Fracturing and the Impact on Water: A Strategic Analysis of the Feasibility of Voluntary Standards," Unpublished Draft, 2015, p. 3.

33 Evan Hansen, Dustin Mulvaney, Meghan Betcher, *Water Resource Reporting and Water Footprint from Marcellus Shale Development in West Virginia and Pennsylvania*, Downstream Strategies, October 2013, pp. viii–ix.

34 Hansen et al., p. viii.

35 Erik Mielke, Laura Diaz Anadon, and Venkatesh Narayanamurti, *Water Consumption of Energy Resource Extraction, Processing and Conversion*, Belfer Center for Science and International Affairs, Discussion Paper #2010–15; and Jesse Jenkins, "Energy Facts: How Much Water Does Fracking for Shale Gas Consume?" *The Energy Collective*, April 6, 2013, http://theenergycollective.com/jessejenkins/205481/friday-energy-facts-how-much-water-does-fracking-shale-gas-consume (accessed July 14, 2014).

36 Reed et al., p. 3.

37 Eliza Griswold, "The Fracturing of Pennsylvania," *New York Times Magazine*, November 17, 2011.

38 Reed et al., p. 8.

39 EPA, *Evaluation of Impacts to Underground Sources of Drinking Water by Hydraulic Fracturing of Coalbed Methane Reservoirs*, June 2004; for a critique of EPA's report, see "Buried Secrets: Is Natural Gas Drilling Endangering U.S. Water Supplies?" *ProPublica, Fracking*, November 13, 2008, www.propublica.org/article/buried-secrets-is-natural-gas-drilling-endangering-us-water-supplies-1113 (accessed July 29, 2014).

40 Reed et al., p. 8.

41 EPA, *Study of the Potential Impacts of Hydraulic Fracturing on Drinking Water Resources: Progress Report*, December 2012.

42 Alex Trembath, Max Luke, Michael Shellenberger, and Ted Nordhaus, *Coal Killer: How Natural Gas Fuels the Clean Energy Revolution*, Oakland, CA: The Breakthrough Institute, June 2013.

43 Ramon Alvarez, Stephen W. Pacala, James J. Winebrake, et al., "Greater Focus Needed on Methane Leakage from Natural Gas Infrastructure," *Proceedings of the National Academy of Sciences*, February 13, 2012.

44 EPA, *Draft Inventory of U.S. Greenhouse Gas Emissions and Sinks: 1990–2013*, 2015; and Anna Karion, Colm Sweeney, Gabrielle Patron, et al. "Methane emissions estimate from airborne measurements over a western United States natural gas field," *Geophysical Research Letters*, 2013.

45 Robert Howarth, Renee Santoro, and Anthony Ingraffea, "Methane and the greenhouse-gas footprint of natural gas from shale formations: a letter," *Climatic Change*, 106, 2011; Howarth, "A bridge to nowhere: methane emissions and the greenhouse gas footprint of natural gas," *Energy Science & Engineering*, 2(2), 2013.

46 David Allen, Vincent M. Torres, James Thomas, et al., "Measurements of methane emissions at natural gas production sites in the United States," *Proceedings of the National Academy of Sciences*, October 29, 2013.

47 Richard Lovett, "Study Revises Estimate of Methane Leaks from U.S. Fracking Fields," *Nature*, September 13, 2013.

48 A.R. Brandt, G. A. Heath, E. A. Kort, et al. "Methane Leaks from North American Natural Gas System," *Science*, February 14, 2014; Dana Caulton, Paul B. Shepson, Renee L. Santoro, et al., "Toward a Better Understanding and Quantification of Methane Emissions from Shale Gas Development," *Proceedings of the National Academy of Sciences*, April 15, 2014.

49 *Our Drinking Water at Risk: What EPA and the Oil and Gas Industry Don't Want Us to Know about Hydraulic Fracturing*, Oil and Gas Accountability Project, 2005.

50 US EPA, "Regulation of Hydraulic Fracturing Under the Safe Drinking Water Act," http://water.epa.gov/type/groundwater/uic/class2/hydraulicfracturing/wells_hydroreg.cfm (accessed July 19, 2014).

51 National Energy Policy Development Group, *National Energy Policy*, May 2001, p. x.

52 US EPA, "Regulatory Determination for Oil and Gas and Geothermal Exploration, Development and Production Wastes," July 6, 1988, www.epa.gov/osw/nonhaz/industrial/special/oil/og88wp.pdf (accessed July 25, 2014).

53 US EPA, *Exemption of Oil and Gas Exploration and Production Wastes from Federal Hazardous Waste Regulations*, 2002, p. 5.

54 David Spence, "Fracking Regulations: Is Federal Hydraulic Fracturing Regulation Around the Corner?" Energy Management Brief, UT Austin Energy Management and Innovation Center, September 22, 2010, www.mccombs.utexas.edu/~/media/Files/MSB/Centers/EMIC/EMIC%20Misc/Fracking-Regulations-Is-Federal-Hydraulic-Fracturing-Regulation-Around-Corner.PDF (accessed July 25, 2014).

55 Spence, 2010.

56 National Resources Defense Council, "State Hydraulic Fracturing Disclosure Rules and Enforcement: A Comparison," NRDC Issue Brief, July 2012.

57 Environmental Defense Center, "Fracking: Federal Law Loopholes and Exemptions," www.edcnet.org/learn/current_cases/fracking/federal_law_loopholes.html (accessed July 19, 2014).

58 Environmental Defense Center, Fracking.

59 US Government Accountability Office, *EPA Program to Protect Underground Sources from Injection of Fluids Associated With Oil and Gas Production Needs Improvement*, June 2014.

60 "Billionaire Father of Fracking Says Government Must Step Up Regulation," *Forbes*, July 19, 2012.

61 US Department of State, "Keystone XL Pipeline Project," http://keystonepipeline-xl.state.gov/archive/c51958.htm (accessed July 14, 2014).

62 Walter Rosenbaum, *American Energy*, CQ Press, 2014, p. 74.

63 US Department of State, *Final Supplemental Environmental Impact Statement for the Keystone XL Pipeline Project*, Executive Summary, January 2014.

64 Elizabeth Kolbert, "A Delay Worth Celebrating? Obama Prolongs Keystone XL Fight," *The New Yorker*, April 25, 2014; KC Golden, "The Keystone Principle: Stop Making Climate Disruption Worse," *Huffington Post*, January 29, 2014, www.huffingtonpost.com/kc-golden/the-keystone-principle_b_4690081.html (accessed April 26, 2015).

65 US Department of State, *Final Supplemental Environmental Impact Statement for the Keystone XL Project*, Chapter 4, January 2014.

66 The National Academy of Sciences, "What you Need to Know about Energy, Coal," http://needtoknow.nas.edu/energy/energy-sources/fossil-fuels/coal/ (accessed May 6, 2014).

67 US Energy Information Administration (EIA), *Annual Energy Outlook 2014*; and EIA, *Monthly Energy Review*, April 2014, p. 95.

68 EIA, *Annual Energy Outlook 2010*; and EIA, *Monthly Energy Review*, August 2015, p. 107.

69 EIA, *Monthly Energy Review*, August 2015, p. 107.

70 US Environmental Protection Agency (EPA), "Regulatory Actions, Final Mercury and Air Toxics Standards (MATS) for Power Plants," www.epa.gov/airquality/powerpla nttoxics/actions.html (accessed May 6, 2014).

71 EPA, "Cross-State Air Pollution Rule," www.epa.gov/airtransport/CSAPR/ (accessed May 6, 2014).

72 EPA, "New Source Review," www.epa.gov/nsr/ (accessed May 6, 2014).

73 EIA, "Coal Plants without Scrubbers Account for a Majority of U.S. SO2 Emissions," *Today in Energy*, December 21, 2011, www.eia.gov/todayinenergy/detail.cfm?id=4410 (accessed May 6, 2014).

74 Federal Register, "Standards of Performance for Greenhouse Gas Emissions from New Stationary Sources: Electric Utility Generating Units," proposed January 8, 2014, www.federalregister.gov/articles/2014/01/08/2013-28668/standards-of-performa nce-for-greenhouse-gas-emissions-from-new-stationary-sources-electric-utility (accessed May 6, 2014).

75 For a good explanation of the proposed rule, see The Center for Climate and Energy Solutions, "EPA Regulation of Greenhouse Gas Emissions from New Power Plants," www.c2es.org/federal/executive/epa/ghg-standards-for-new-power-plants (accessed May 6, 2014).

76 EPA, "Carbon Pollution Emission Guidelines for Existing Stationary Sources: Electric Utility Generating Units, Final Rule," www3.epa.gov/airquality/cpp/cpp-final-rule. pdf (accessed October 10, 2015).

77 EPA, "Carbon Pollution ...., Final Rule," pp. 1,556–7, www3.epa.gov/airquality/cpp/ cpp-final-rule.pdf (accessed October 10, 2015).

78 "Obama Seeks 30 Percent Cuts in Power Plants' Carbon Pollution," *Politico*, June 2, 2014, www.politico.com/story/2014/06/reports-obama-to-seek-30-percent-cuts-in-p ower-plants-carbon-pollution-107301.html (accessed July 9, 2014).

79 EPA, "Regulatory Impact Analysis ....," p. ES-24.

80 Executive Office of the President, *The President's Climate Action Plan*, June 2013, p. 19.

81 *SCOTUS Blog*, "Utility Air Regulatory Group v. Environmental Protection Agency," www.scotusblog.com/case-files/cases/utility-air-regulatory-group-v-environmental-protec tion- agency/ (accessed May 9, 2014).

82 Senator Mitch McConnell, *Letter to the National Governors Association*, March 19, 2015.

83 Jeff McMahon, "States Ignoring Mitch McConnell, Working on Clean Power Plan: EPA," *Forbes*, April 12, 2015, www.forbes.com/sites/jeffmcmahon/2015/04/12/sta tes-ignoring-mitch-mcconnell-epa/ (accessed May 1, 2015).

84 EIA, *Annual Energy Outlook 2014*.

85 EIA, "Planned Coal-Fired Power Plant Retirements Continue to Increase," *Today in Energy*, March 20, 2014.

86 US DOE, Office of Fossil Energy, "Enhanced Oil Recovery," http://energy.gov/fe/ science-innovation/oil-gas/enhanced-oil-recovery (accessed May 9, 2014).

87 Global CCS Institute, "Large Scale CCS Projects," www.globalccsinstitute.com/p rojects/large-scale-ccs-projects (accessed May 1, 2015).

88 Global CCS Institute, "How CCS Works – Transport," www.globalccsinstitute.com/ understanding-ccs/how-ccs-works-transport (accessed May 9, 2014).

89 Congressional Research Service (CRS), "Clean Coal Authorizations, Appropriations and Incentives," November 1, 2010, p. 10; and CRS, "Carbon Capture and

Sequestration: Research, Development, and Demonstration at the U.S. Department of Energy," February 10, 2014, p. 8.

90 US Department of Energy, *DOE/NETL Carbon Dioxide Capture and Storage RD&D Roadmap*, December 2010, p. 3.

91 DOE, *Roadmap*, p. 10.

92 CRS, "Carbon Capture and Sequestration," pp. 10–11.

93 DOE, *Roadmap*, p. 11.

94 CRS, "Carbon Capture and Sequestration," pp. 13–22.

95 EIA, "Levelized Cost and Levelized Avoided Cost of New Generation Resources," April 17, 2014, www.eia.gov/forecasts/aeo/electricity_generation.cfm (accessed May 9, 2014); Mohammed Al-Juaied and Adam Whitmore, "Realistic Costs of Carbon Capture," Belfer Center Discussion Paper 2009–08, July 2009, p. ii; International Energy Agency, *Cost and Performance of Carbon Dioxide Capture from Power Generation*, 2011, pp. 26–31.

96 Chris Nelder, "Why Carbon Capture and Storage Will Never Pay Off," *ZDNet* (CBS Interactive), March 6, 2013, www.zdnet.com/article/why-carbon-capture-and-storage-will-never-pay-off/ (accessed May 1, 2015).

97 The Energy Collective, "Why We Need CCS, Part 2: Reactive Climate Change Mitigation," May 7, 2014, http://theenergycollective.com/schalk-cloete/373156/why-we-need-ccs-part-2-reactive-climate-change-mitigation (accessed June 12, 2014).

98 CRS, "Carbon Capture and Sequestration," p. 6.

5

DOMESTIC POLICY AND
ENERGY TRANSFORMATION

Renewables

The burgeoning transformation in America's energy profile is happening not only with regard to fossil fuels, but with respect to renewable energy as well. Public policy has been the key driver behind this change, and likely with good reason. Renewable energy resources are rightly understood to hold quite significant potential to help achieve various aspects of energy security. If most or all of America's electricity came from renewables, one could imagine a circumstance in which endless supplies of cost-free "fuel" (think sunlight, wind, heat from underground sources, water in motion) powered the nation's energy needs. Combined with a growing fleet of electric commercial and private vehicles, this scenario could potentially yield numerous benefits such as the following: reduced dependence on oil, especially foreign sources; little or no mining and burning of coal, which has produced the most adverse health and environmental impacts of all the major energy sources; more money being invested and spent domestically on energy development and consumption; less vulnerability to volatile global energy markets and supply chains; and of course, reduced air pollution and fewer CO_2 emissions. As Amory Lovins argues:

> An 80-percent-renewable, half-distributed, nearly decarbonized, highly resilient US grid could cost virtually the same as business as usual, but could best manage its risks—security, technology, finance, climate, health, fuel, and water—and, uniquely, prevent cascading blackouts.[1]

In all, it would be a blessing, a large step toward reaching the goal of achieving energy security. The good news is that over time, it is expected that more of these benefits can be realized as production and use of renewables continue to ramp up in the United States and abroad.

At the same time, like the well-used expression attributed to residents of New England regarding the inadvisability of getting to "there" by starting from "here," the path to a renewable energy-fueled future presents some substantial obstacles along the way. These impediments have been made evident due to the nature of three of the most widely adopted renewable energy sources, all of which have been targeted for significant expansion by public policy—namely, wind and solar in the electric power sector, and ethanol in the transportation and liquid fuels sector. Grafting new technologies onto existing production systems, supply networks, grid infrastructure, business models and regulatory frameworks is prompting major changes to all these components of the US energy system. It can be credibly argued that the transformative benefits of renewable energy described above justify the turbulence inherent in the market changes associated with the adoption of these admittedly disruptive technologies. Nonetheless, it is important to critically examine these changes to more fully understand the many implications and substantial costs of the effort being made to achieve greater energy security through greater reliance upon renewable sources of energy.

Looking at wind and solar power first, and ethanol in the following section, this chapter examines how the realm of public policy has both contributed to the growth of these alternative energy sources, just as the policy measures were intended, while at the same time ushering in an array of unanticipated costs and challenges to further growth in the renewables area. The evolution of wind, solar and ethanol promotion through governmental market inducements demonstrates the nature of the energy security dilemma as encountered in the renewable energy source development realm of US public policy.

Promoting Renewable Electricity

Wind and solar technologies have in recent years captured a great deal of public and political attention, along with business investment and consumption by industrial, commercial and residential end-users. Generous policy incentives and aggressive mandates, at both the federal and state levels, have fueled remarkable growth in these sources of electrical power generation. In 2014 wind power accounted for 6.5% of all generation capacity in the United States, and 4.5% of all electricity consumption. Total installed capacity increased by more than 50% from 2010 through 2014, from 40,000 to almost 66,000 MW.[2] Total installed solar PV capacity in the US surpassed 22,000 MW in 2015, up from 4,000 MW only four years earlier.[3] It is expected that the use of wind and solar, and to a lesser extent other renewables, will continue to increase in the coming years, perhaps as rapidly as has been the case over the last decade. However, at the same time, this growth has led to some largely unintended consequences with which the electric power sector is struggling to cope as they move toward the future.

The promise and challenge of renewable energy, especially wind and solar power, have derived from two overarching aims that have been enshrined in different realms of public policy. These two goals of policy have only recently begun to collide with one another, forcing the kind of tradeoff calculations we have argued are part-and-parcel of achieving the goal of maximal energy security at any point in time.

The first of these areas has been the policy measure that exists in every state in the country, a mandate requiring power utilities to provide electricity for all end-users in their respective territories. Because utilities serve a public interest (regardless of whether they are investor-owned utilities, municipal utilities or rural electric co-ops), they are required to offer and provide service to any customer who requests it and can pay for it at the regulated price. (This can be a market price as opposed to a fixed rate, but such pricing structures are provided for by regulation.) In other words, service is universal. This is sometimes referred to as a "regulatory compact," an agreement in which the utility accepts an obligation to provide power to all users in exchange for the government's promise to set prices that will allow the utility to be compensated fully for the provision of that service. While the universality concept is not codified in the United States as an actual contract between utilities and government entities, in the American legal system it captures the idea inherent in the requirement of a natural monopoly to provide service.[4] This is particularly important with regard to solar power and other forms of distributed generation, as end-users can rely upon their utilities to provide power when their own on-site units are not providing power, either as a result of planned or unplanned service interruptions (i.e., there is no expectation that a solar PV system will provide electricity at night, or the system could fail due to a technical malfunction).

The second realm of public policy that has contributed to both the growth of the renewable energy sector and its obstacles has been the widespread adoption and use of robust financial incentives and statutory/regulatory mandates in federal, state and local jurisdictions to promote the use of renewable energy technologies. With regard to federal policy, these incentives and mandates cover multiple technologies and economic sectors, having been addressed in numerous pieces of legislation. While the first federal efforts to promote solar power and other renewables generally fell flat in the 1970s (see Joshua Green, "The Elusive Green Economy"[5]), a few key provisions were adopted that in hindsight have proven to be rather critical over time. One of these noteworthy provisions was the Public Utilities Regulatory Policy Act (PURPA). PURPA established that qualifying generation facilities not owned by a utility could sell power to utilities. In fact, if independent power producers met the specified qualifications, utilities were required to take the power at the avoided rate (the amount the utility would otherwise have to pay to acquire the electricity). This legislative stipulation became known as the "must take" power provision. While PURPA was proposed in part with renewable energy in mind, it became useful for building markets for a variety of generating facilities, and for ushering in greater restructuring of the electric power sector. Further, it created an important example of privileged status in establishing renewable energy generation as a special class of electricity generation that would receive preferential treatment by law. Subsequent legislation in the *Energy Policy Act of 1992* further opened up the wholesale market for independent power producers, allowing for faster growth of renewable generation. In addition, numerous states and other utility jurisdictions have imposed similar requirements on utilities to take all power emanating from renewable energy generation facilities. This has been especially important with regard to the growth of wind power, and the efforts by regulators to ensure that energy from wind is integrated into the nation's power grid as much as possible.

By the 1990s and early 2000s, when the federal government began to make renewed efforts to advance clean energy sources, the times, the technologies and the circumstances had changed since the 1970s when these forms of energy generation were first proposed. The growing concern with global climate change, the increasing output and cost-effectiveness of solar and wind technologies, and the involvement in two wars with Iraq (which highlighted the issue of energy security), all combined to prompt a new critical mass of support to develop domestic clean energy sources while promoting energy efficiency. In this context, the combination of advocacy by domestic industry representatives, along with the growing number of organizations and individuals dedicated to environmental protection and sustainability, allowed for the adoption of federal policy, particularly in the form of sizeable tax incentives to promote the expansion of the renewable energy sector.

The two major federal tax incentives that have been used the most frequently are the Investment Tax Credit (ITC) and the Production Tax Credit (PTC). The ITC allows commercial and residential purchasers of qualifying technologies (solar PV, solar water and space heating, small wind turbines, ocean tidal and wave, geothermal, fuel cells, and several others) to claim a tax credit totaling a specified percentage of the cost to install the energy generation system. The installation of solar technologies, particularly PV, constitutes by far the largest use of the ITC. The ITC was increased from 10% to 30% with the *Energy Policy Act of 2005*, and has remained at this level since then (the level for geothermal, microturbines and combined heat and power has stayed at 10%). The ITC serves to build markets by incentivizing the production and use of technologies that otherwise might not be cost-effective to install. It has been particularly effective in the development of on-site generation systems (those built at the site where the energy will be used). For example, on the larger side, Toyota has installed large PV systems at some of its facilities in the United States (two of them are more than 1.5 MW), while at a smaller level, PV systems, usually about 3–5 kilowatts, have been installed on hundreds of thousands of homes.[6] The ITC has also been increasingly essential to the development of large, utility-scale projects. New Mexico, for example, has seen an increase in the installation of such projects, including the 30 MW Cimmaron Solar Facility and the 50 MW Macho Springs Solar Project.[7] The ITC was set to revert to 10% at the end of 2016, and while a strong coalition secured a five-year extension from Congress, further extensions are by no means certain. The irony is that the very success of the ITC in encouraging solar adoption may spell its ultimate demise, as it has made the incentive more expensive to maintain, and less necessary to support the industry which is now beyond the start-up stage of operation.

The PTC works differently from the ITC. Instead of providing a tax credit based on the costs associated with an energy system, it provides a tax credit based on the actual energy production of the system in question. This has been the incentive used to fuel the rapid growth of the market for utility-scale wind turbines, though it is also applicable for other renewable energy technologies (biomass, ocean, landfill gas, geothermal). The PTC was adopted in the *Energy Policy Act of 1992*, and as with the ITC it has been extended several times for a few years at a time (though unlike the ITC, the PTC has been allowed to lapse several times since its adoption). The PTC in 2014 provided an incentive of 2.3 cents per kilowatt-hour for energy generated over the first 10 years of operation of a qualifying system. The incentive started at a level of 1.5 cents/kWh in 1993, and has since been indexed for inflation. The PTC expired at the end of 2014 (it had expired at the end of 2013, but was extended retroactively by Congress at the end of 2014), but Congress later extended it through 2019 amidst uncertain political support. It may very well be the case that the very success of the PTC has made the incentive more costly, more controversial, and, perhaps no longer necessary to assure continuance of the industry.

The PTC has been useful for developers of large renewable energy systems, as they can sell and track sufficient volumes of electricity to make a production-based incentive workable (trying to provide a production-based incentive for numerous smaller, on-site systems makes little sense; it's much easier to provide a one-time up front incentive such as the ITC instead of having users track annual production and claim it on their tax returns over a period of several years). The wind power industry has been able to take great advantage of this particular incentive—and lobbies heavily for it—because wind resources are in fact relatively abundant in the United States, allowing turbines to be installed throughout the entire country (the central part of the US, from North Dakota down to the Texas panhandle, enjoys such strong wind resources it has prompted references to the US as being "the Saudi Arabia of wind"). In addition, the wind industry's extensive use of the PTC is attributable to the fact that the turbine technology at the core of the industry is relatively mature, especially in comparison to other newer renewable energy technologies such as wave, tidal and geothermal. Nonetheless, even though the technology is well-established, the size and output capacity of wind turbines continues to grow as the industry has matured both in the United States and in many countries abroad. Turbines now commonly in use have a capacity of 2 MW or more, with blades up to 60 meters in length, and towers extending 100 meters or more into the air. (The largest land-based turbine, which is an anomaly at 7.5 MW, is located in Germany, while a handful of

5–6 MW turbines have been deployed offshore in Europe. These very large turbines have generally been constructed for test and demonstration purposes, and are not available for widespread use.[8])

Federal policy has played a key role in making the wind and solar power industries successful in this country. By using both the ITC and PTC, the federal government has been able to enact and maintain policies that are acceptable to the public and the energy industry alike, while spurring the growth of clean energy industries whose success can mitigate the growing environmental liabilities arising from the continued large-scale use of fossil fuels.

Alongside federal policy, several US states have made an aggressive effort to promote the use of both wind and solar power, and these policies have also had important effects on the growth of these types of renewable energy sources. California is the acknowledged leader in solar energy due to its generous incentives, while Texas is the largest wind producer in the country as a result of its favorable policy environment. Consistent with the idea that the states act as public policy "laboratories" to test different types of policies, the variety of policy tools used at the state level has been far greater than those few tax advantages and pilot project funding initiatives employed at the federal level. California, for example, uses a production-based incentive for large solar installations and an investment-based incentive for smaller ones (though to remain consistent with the idea of having a production-based program, California has called its incentive for small systems an "Expected Performance-Based Buydown"). The incentive levels, which are very generous compared to many other state-level policies elsewhere in the country, will go down in value for future solar adopters, as more solar power gets installed in the state. In other words, there are incentives to being early innovators; those who install solar sooner will enjoy bigger benefits than those who follow later on.[9]

Many other states also provide their own incentives in the form of tax credits or rebates (such as New Mexico, which accounts for the two large solar projects cited above). These incentives can be combined with the federal ITC to make the purchase of solar energy even more cost-effective, allowing a system to pay for itself over a period as short as 5–7 years (sometimes the payback is even quicker, but it's usually longer). Some states have also established dedicated funds and organizations to support renewable energy and energy efficiency-promoting practices. The Energy Trust of Oregon, for example, is funded by a 3% surcharge on consumers' utility bills. The Energy Trust provides support for a variety of renewable energy and conservation activities, including on-site solar PV, providing homeowners a cash rebate of up to 70 cents per installed watt, capped at $7,000, and commercial customers anywhere from 38 cents to $1.10 per watt, capped at either $76,000 or $150,000, depending upon the utility provider and the system size.[10] The impact of these types of policies is clearly demonstrated by the fact that states that provide incentives for solar are the ones that have seen the greatest adoption of solar power, in spite of their solar resources. For example, states in the southeastern United States, which have plenty of sun but offer comparatively less support for solar power, have relatively low adoption rates. By contrast, Massachusetts, New Jersey and for a time, Oregon—places not particularly known for their sunny, warm climates—have been among the leading states for solar installations (in Oregon, the sizable 50% ITC for businesses ended in 2011, dropping the state out of the list of national leaders since that time).[11]

Another noteworthy policy measure, which has been adopted in 38 US states, is the Renewable Portfolio Standard (RPS). Unlike a tax credit or rebate to incentivize changes in energy use, an RPS involves a mandate that a specified level of electricity within a jurisdiction must be generated from renewable resources by a particular year (for example, Illinois calls for 25% of its electricity to come from renewables by 2025, with the target level rising in small increments annually until such time as the goal is met). Utilities not in compliance with the annual targets are required to pay a fine based on the amount they fall short. Some states, such as New Jersey and Massachusetts, include a specified solar power target, with very high non-compliance payments. For example, in Massachusetts the solar power non-compliance payment for utilities was 52.3 cents per kilowatt-hour in 2014. Since electricity sold in the state at 12–16 cents per kWh, and utilities could purchase solar energy (technically, they purchased solar renewable energy credits, or SRECs) for 25–30 cents per kWh, the utilities subject to the RPS have had a great incentive to acquire solar-generated electricity.[12] Texas has taken a different route to renewable energy development, focusing its attention on wind power. Its RPS calls for 10,000 MW of renewable capacity by 2025. As of 2013, the state already had more than 12,000 MW of wind installed; it reached its goal 15 years ahead of schedule.[13] In addition, Texas created "Competitive Renewable Energy Zones," a policy by which the public utilities commission is permitted to designate certain areas where wind generation facilities would be constructed, along with the transmission lines necessary to move the electricity to consumers.[14] Target levels, non-compliance payments, and other terms and conditions vary by state, but overall RPS mandates have greatly contributed to the growth of the renewable energy sector in numerous states in the United States.

These two broad, overarching aims—ensuring the availability of electric power to consumers, and expanding the production and use of renewables—have been pursued largely independently of one another. They have resulted from separate public policy processes, often occurring in different "action arenas," at different times, involving different "advocacy coalitions" supporting these very different types of goals. For a long time there was little or no contradiction between these two objectives. In the early years of developing technologies and markets for wind and solar power, the scale of their use was too small to have an important impact on the US electric power sector. However, as the adoption of wind and solar technologies has greatly increased, so too has the conflict between these two policy realms become more evident. The problem derives from two particular qualities of wind and solar technologies and how they have been deployed—namely, the fact that they are both intermittent resources, and the fact that they are distributed widely throughout the nation's established electric power grid.

Handling (Too Much?) of a Good Thing

The Variable Generation of Intermittent Resources

The "fuels" for wind and solar power—sunlight and wind, respectively—are free. They are both widespread and relatively abundant. They don't pollute, and they don't run out over time. These are good things, and as Mae West famously pointed out, "Too much of a good thing can be wonderful." However, solar and wind power are not available at all times, so managing the transition toward their expanded use is indeed critical. The timing and strength of the wind cannot be predicted beyond averages spread out over time, and while the timing of sunlight is known, its strength is not (due to unpredictable levels of cloud cover). This variable generation has implications for energy security, and for energy source reliability, and it presents a potential problem for the powerful utilities and regional transmission operators (RTOs) whose job it is to provide continuous electric power to end-users. This is because production and consumption must be balanced in real time so that they are equal. (The utilities and RTOs that have this responsibility are known as balancing authorities.) Otherwise the grid can experience blackouts, brownouts, cascading outages, or other types of failures. When electricity is supplied by coal, natural gas, hydro or nuclear power plants, operators can predict precisely how much electricity they will have to supply at any given time—they can control this ratio with great precision and reliability. Accordingly, balancing authorities can schedule production to meet expected consumption with well-established practices and known technologies. Exact consumption levels are themselves somewhat unpredictable. Demand can and does sometimes fluctuate greatly in a short time. However, copious data exists—on an hour-by-hour basis for every hour of the year—on average consumption levels in utility territories throughout the country. Having reliable information about supply and demand makes balancing a more manageable task (other information sources such as weather forecasts and economic data, along with backup reserve power, are also key elements to keeping the grid in balance). When a large enough supply of intermittent resources are added to the mix, and currently in the US this primarily means wind power, the supply side of the equation can change quickly, leaving the grid either under- or oversupplied at any one point in time. The combination of both demand fluctuation and variable generation complicates the management of the grid.

Figure 5.1 shows the level of output from wind energy and other resources that have to be balanced by the Bonneville Power Administration (BPA) in the Pacific Northwest region of the United States. Wind is shown on the lowest line,

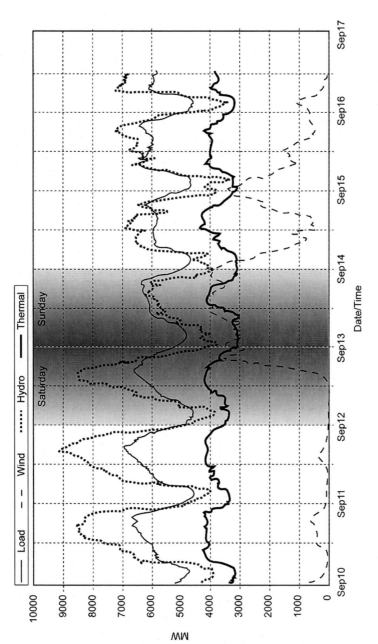

FIGURE 5.1 Bonneville Power Administration—Electricity Supply by Source, September 2015
Source: Bonneville Power Administration.

and its output varies a great deal without any easily predictable pattern.[15] Since the output from wind can change rapidly, it greatly complicates the job of balancing grid supply with demand. The problem can be exacerbated in cases where the balancing authority is required to take power from the intermittent suppliers as a top priority.

A few examples can illustrate the nature of the problem being confronted by power grid operators and public utilities. In the Pacific Northwest in 2011, when the BPA had too much supply to meet demand during a given period, it told wind suppliers in the region that it would not take their power. This decision caused a great stir, particularly among proponents of wind energy, as BPA had contracted with the wind farms to take their power. However, the BPA is also required to protect endangered salmon in the Columbia River, and at the time these two obligations conflicted—in the way so many aspects of energy security entail difficult tradeoffs. As the newspaper *The Oregonian* put it, "When the spring snowmelt is so profuse that river water must be stored or spilled over dams instead of through power generating turbines, dissolved gases in the river rise— potentially a mortal threat to salmon. So at peak moments of electricity production, it's hydropower or wind—and limiting wind might save fish."[16]

Even without the competing environmental values that added complications to the issue involving the BPA, the problem of balancing intermittent resources is growing in significance and frequency. When variable generation and fluctuating demand lead to an oversupply of electricity, one result can be negative pricing. For example, in Texas the oversupply of wind energy has caused prices to fall below zero multiple times. In other words, generators have found themselves in a position of paying other entities to take their power. While the installation of added transmission capacity in Texas since 2011 has diminished this occurrence (more electricity generated in windy West Texas could be moved to the population centers in the eastern part of the state), the high level of wind power in Texas can still occasionally lead to negative pricing. In February 2015, wind energy generation one night reached a record 11,154 MW, accounting for 41% of all electricity generated in the state. That record was broken again a few months later, and throughout this period, prices fell below $0/MWh several times.[17] A notable example of this problem occurred in June 2013 in Germany. An oversupply of electricity, with roughly half of the country's needs (45,000 MW) being provided one day by relatively new solar sources, caused the wholesale price of electricity to fall to negative 100 Euros/MW, resulting in the loss of 500 billion Euros. In cases such as these it can still pay to generate power even if the price of electricity falls below zero. If subsidies or other incentives or payments to generators exceed the amount of the financial loss, it can still make sense to produce the electricity rather than shut down.[18] This circumstance, resulting from unpredictable variable generation, points to a potentially serious problem in electricity markets as more power comes to be generated with renewable sources.

There are three types of solutions that currently dominate the discussion of how to address the growing problems of over- and under-supply of electricity. One solution is simply to build more backup generation, at least to handle undersupply. Natural gas plants offer a good form of backup because they can be ramped up quickly, and they are attractive in the United States during this time of low natural gas prices. But backup power still needs to be cost-effective, as it competes against the renewables it would also back up. The CEO of California's Independent System Operator succinctly characterized the issue, stating that natural gas plant operators "[are] not getting as much revenue as they once did because they're not selling as much power because it's being displaced by wind and solar energy, which is exactly what we want."[19] Nonetheless, considering the way that the grid is managed, in which all power generation is backed up in order to ensure service in the case of disruptions (utilities and RTOs are required to schedule sufficient levels of backup power and equitably allocate costs in case of unplanned outages), it should be possible to do the same for growing levels of intermittent resources.

A second idea is to develop a more sophisticated system to manage the power grid. This type of system is one of the technological elements of what is now called the "smart grid." The term smart grid is a catchall term that refers to the introduction of new technologies at various points in the operation of the grid—ranging from power plants to single on/off switches on individual appliances in homes and businesses—that can at all times monitor use, communicate information to advanced meters, generators, grid managers and consumers, and direct or cut off the flow of electricity to particular places. If the right software and appropriate hardware can be developed so that balancing authorities can manage the grid and balance all generation assets (baseload, peak, distributed, intermittent, etc.) with all the consumer demands that are met by various utilities and service providers, the problem could more readily be addressed. These kinds of smart grid technologies are being researched, developed, and demonstrated around the country in collaborations among utility companies, government, universities and end-users, often with funding from the federal government. However, despite some progress in the development of the software and hardware required, and the installation of millions of "smart meters" around the country, such integrated systems have been created only in a few small-scale test sites.

Tied to the idea of building a smarter grid is creating smarter consumers through pricing mechanisms such as "time-of-use" rates (also called "dynamic pricing"), and the implementation of other "demand response" measures. When more information can be provided to consumers about their choices and the associated costs, they can be expected to change their behavior in ways that can significantly affect the total demand of electricity at given times. While commercial and industrial customers pay time-of-use rates that rise during peak hours of demand, residential customers do not, and so they tend to have little knowledge of how their electricity usage affects the management of the grid. If residential customers paid a higher rate to use electricity during peak hours, they would be more likely to curtail certain uses until the price of electricity went back down (i.e., washing and drying clothes at night instead of in the afternoon). While this would make the grid more manageable and diminish imbalances and price spikes, it would also be a tough sell to many consumers, who would be likely to see their electricity bills increase if they persisted in their established patterns of use. Public utilities commissions asked to adopt time-of-use rates for residential customers would be likely to face great opposition from consumer advocates and politicians alike.

A more palatable type of approach to managing supply and demand of electricity involves other demand response and management measures. The idea behind demand-side responses is to alter the timing of demand, reduce peak loads, and/or cut overall consumption through automation and/or financial incentives. Carried out on a large enough scale, demand management could preclude the need to build additional generation capacity for heightened demands that may occur only a few days, or even a limited number of hours, in a typical year. Some of these measures can be technological, allowing for some uses to be curtailed automatically. For example, smart grid technologies could be employed to automate demand response. They could communicate with the utility to enable grid managers to reduce or turn off power for a period of time to heating or air conditioning units, water heaters, and refrigerators. On a hot summer day, having everyone in a big city such as Los Angeles live with the air conditioning a couple degrees higher, or cycling off their refrigerators for a half hour, can have a major impact on peak load. Other measures are behavioral, requiring consumers to make choices and take appropriate action at different times of the day. Providing real-time energy use information, either through a smartphone app, a smart meter, or an "energy dashboard" (all of these products are currently available) can potentially alter people's behavior favorably and prompt them to use less electricity. In order to achieve these lofty goals, however, a great deal of time, investment, and coordination will be needed on the part of industry, regulators, policymakers and consumers. The effort to do this is in fact well under way. Initiatives such as FutureGrid and GridWise receive federal funding through the US Department of Energy to try to develop the technologies and management systems that will enable large-scale demand response programs to be operating throughout the grid. State programs such as California's Demand Response Research Center have also been established through legislative action to advance the realm of demand response.

A third approach—really one part of the smart grid realm—is to store excess electricity in times of oversupply, and then use it in times of undersupply. If large amounts of electricity could be stored, and then used whenever they have to be called upon, the balancing problem posed by intermittent resources would largely disappear. However, electricity supplied to the grid is not capable of being stored at a large-scale, and until cost-effective technologies become available to do this, storage will not be an option available for managing the grid. Industry is responding to the need. While pumped storage has been the most used technology to date, its use is not widespread. Batteries appear to be the technology that has captured the most attention and investment, and research and development in this sector has brought costs down significantly in recent years. Tesla, known for its electric cars, unveiled in 2015 its "Powerwall" battery for home energy storage, and these 7kWh or 10kWh systems can be combined to meet larger needs. Compressed air offers another opportunity to eventually achieve utility scale energy storage. These innovative technologies provide great promise. However, when it comes to integrating storage technologies into the grid, hopes and expectations far exceed actual results.

It may be possible that increasing the supply of variable generation resources from an even greater variety of generators would solve the problem of intermittency, as long as enough transmission lines are built to move large amounts of electricity. One study from 2013 suggested that it would be possible to have reliable supplies in a scenario where 99% of the US electricity demand could be provided by 2030 using wind and solar, along with some strategic storage capacity.[20] This understanding is based on the idea that reliability, and therefore energy security, is in part a function of diversity of sources. With wind and solar, such diversity might be most effectively approached by geographical diversity—siting generation systems in a large number of interconnected sites so that there will always be an expectation of a minimum level of electricity produced collectively. The researchers in this study considered 28 billion combinations of renewables and storage, tested with four years of load and weather data, and assumed the use of currently available technologies. This type of potential solution seems to suggest that the important problem of intermittency can and will take care of itself (sometimes quantity has a quality all its own), contradicting the ongoing concerns of grid managers. However, storage does not yet seem to be a realistic answer sufficient to meet the scale of the challenge, as it does not offer a viable solution in the short- to medium-term timeframe.

As always, the policy realm is part of the effort to find a solution. With regard to smart grid and energy storage, federal funding in the *American Reinvestment and Recovery Act of 2009* (ARRA) included a sizable infusion of money to advance these efforts, providing $4.5 billion to hundreds of public utilities, industry organizations, municipalities and others for research, development and demonstration projects. Billions of dollars have been spent on advanced meters, grid management software and hardware, communications management systems, and electricity storage technologies, to name a few.[21] The Federal Energy Regulatory Commission (FERC), in another example, is helping to facilitate the deployment of storage technology. In 2013 FERC issued Order 792, which adds energy storage as a power source that is eligible to connect to the grid under the same terms as other small electricity generators, including a "fast track" process.[22]

Among the states, California is again ahead of all others, and has taken an interesting approach to the problem of electricity balancing: it has mandated the use of storage technologies. The state's Public Utilities Commission adopted a measure in 2013 that requires investor-owned utilities to buy and install electricity storage technology, reaching a total of 1,325 MW of capacity by the end of 2020.[23] The logic is that the requirement will prompt the needed investment in technology development and subsequent deployment. (When potential investors ask technology companies why utilities and other power suppliers would want to buy their products, which may be quite costly, the answer will simply be, "Because they are required to buy it.") This is most certainly a risky approach, and the success of a policy "solution" that requires an unproven technological answer to a technical problem arising out of a particular business model remains to be seen. The California Independent System Operator has also created an "energy imbalance market," by working with balancing authorities in several other Western states to integrate resources more efficiently and diminish the problem of intermittency. This real-time market, which began in 2014, is expected to produce less volatility and greater reliability with regard to the availability of generation resources and their prices.[24] An interesting example of this imbalance involves the 2–3 hour window around sunset each day, when solar panels stop producing but wind turbines are not yet operating at sufficient capacity to replace their level of production (the wind power usually peaks at nighttime).[25]

In addition to supplying funds and enacting mandates to implement technical solutions to intermittent generation, there are other policy measures that offer some ways to address this problem. One such measure is to relax requirements that generation from wind and solar resources be given priority by grid managers. In states that privilege renewable resources with "must take" requirements, utilities and balancing authorities are mandated to take all electricity produced from wind and solar generators. Utilities and grid managers have recommended that they be able to curtail intermittent generation during periods of low demand or oversupply.[26] However, considering the popularity of renewable energy and the great efforts made to expand its use, there seems to be little public appetite to provide too much latitude in reducing wind and solar usage.

Distributed Generation

A second complicating factor in the effort to expand the production and use of renewable energy is that such technologies, particularly solar PV, are often widely distributed in relatively small systems. Distributed generation (DG) is a term that refers to the siting of electricity generation systems at the locations where the energy will be used, though other terms such as on-site generation and distributed energy resources are commonly used as well. Unlike conventional power plants, which are large in size but small in number, DG systems are, by contrast, small in size but large in number. In the United States, the number of DG systems installed over the past decade has grown at a rapid rate. Far outpacing the installation of all other types of systems has been solar PV, which has grown from 83 MW of total capacity in 2003 to over 22,000 MW in 2015.[27] California is the hands-down leader in solar DG in the United States, followed by Arizona, New Jersey and North Carolina.[28] (Overseas, Germany boasts the greatest adoption of solar PV, followed by Italy, Japan, China and Spain; this is based on both installed capacity and actual generation, though the order of the list changes depending upon which metric is used.[29]) As has already been mentioned, the growth of this market has been driven in large part by public policy, with the federal ITC, along with a variety of state-based measures, accounting for its rapid expansion. At the same time, support for the solar industry via policy (not only in the United States, but also in Europe, Japan and China) has helped to spur the market, contributing to increased production, technology advances, and noteworthy price reductions over time. The total amount of worldwide capacity at the end of 2014 was at 177 GW, with more than two-thirds of that amount being installed after 2010.[30] Moreover, since the 1970s when federal and state programs in the United States began providing support for solar energy, the cost of solar panels has come down more than 99%, going from roughly 75 dollars per watt to less than 75 cents per watt (this is only for the panels, and does not include installation costs, which are higher than the panels), allowing solar power to get ever closer to achieving "grid parity."[31] One could argue that the incentive system is working as planned—that is, a promising technology has received government support, allowing the industry to grow and thrive, bringing costs down and in due course allowing the industry to stand on its own. While this last part has not yet happened, and there are strong differences of opinion regarding whether solar can ever be successful without government subsidies, there is a compelling case to be made for the progress of solar power. (Remember too, that tax incentives and subsidies still support the oil and gas industries, as well as coal, nuclear, and hydropower. No current energy source exists without government support; this is perhaps the strongest evidence of how very important energy security is for any nation.)

On-site generation, particularly PV, can provide a number of benefits to those who install such a system. For homeowners, it appeals to a desire to promote environmental stewardship by providing energy from a clean, renewable source, while promising greater self-sufficiency by diminishing reliance on electricity from utilities (which sometimes do not have a great reputation among the public and businesses they serve). Most utilities rely heavily on electricity produced from fossil fuels, and can readily be depicted as highly profitable corporations that earn millions of dollars while polluting the air and causing global warming. In addition, on-site generation reduces the need for transmission lines, and saves energy by eliminating line losses incurred by the transmission of electricity across long distances (though admittedly, there are probably few residential consumers thinking about this as a factor in their purchase of PV).

Companies installing on-site PV systems also receive benefits. They can help to meet corporate environmental or greenhouse gas emissions goals, and many large multi-national corporations have such goals, as part of Corporate Social Responsibility efforts or other programs.[32] On-site generation can add to energy security, ensuring at least some portion of electricity supply during operating hours. It can also help to manage the volatility of energy costs. When a PV system is installed, both the cost of the system and the expected average output are known quantities, so energy users will know their costs over the next several years with regard to some portion of their electric bill. This information offers protection against both steadily increasing utility rates over time, and against high time-of-use rates that can hit particularly hard during times of peak demand. Rates in the afternoons tend to be the highest, just when a PV system will produce its highest output, so PV can replace some of the power provided by the power utility. This type of "peak shaving" can offer important savings to those large companies, private residences, and small businesses that have high electricity demands.

As distributed generation has proliferated, so too have its impacts been more widely noted, such that the use of on-site solar PV at ever-increasing levels has started to present some problems for managing the power grid. Utilities and grid operators have identified a few key issues of concern, both technical and financial in nature. The major technical concern involves the integration of electricity produced by DG systems into the grid. The larger part of this challenge goes hand-in-hand with solving the problem of integrating resources from large wind farms—namely, intermittency. On-site solar PV is produced on an unpredictable schedule, and balancing the supply with the load from disparate PV systems whose output can fluctuate in a matter of minutes offers an enormous technical challenge for power grid management. A second integration challenge involves the complexity of integrating this off-grid network power output—often at very low levels—from multiple sources (hundreds of thousands in a place like California). This problem derives from the fact that DG systems remain fully connected to the grid, since they do not provide all the electricity needed on-site at all times. Sometimes they produce less than is needed, sometimes more. When they produce more than is needed (e.g., homes during the day when people are at work, businesses closed on the weekend), the electricity is supplied to the grid. Integrating these resources at high levels of market penetration can be a complex task; the challenges involve grid instability, synchronization, "islanding" effects, interconnection standards, active power, reactive power and voltage support.[33] When grid operators are dealing with a limited number of central power plants and big wind farms, the technical elements of integrating supplies are less complex than is the case when dealing with myriad DG systems. What all this DG capacity means is that the old, top-down structure of the grid is being replaced by a more complex system increasingly characterized by a bi-directional, continuously fluctuating flow of power.

A second consequence that is arising from greater DG usage is a purely financial one: the utility industry sees that the rapid growth of DG threatens the continuation of its long-established business model. Market actors in the United States can point to Germany as a harbinger of the future, as utilities *E.on* and *RWE* have seen corporate revenues and stock prices fall in the wake of the substantial market penetration of solar PV.[34] Looking at the US market, *NRG Energy* CEO David Crane observed in 2013 that utilities "realize that distributed solar is a mortal threat to their business."[35] This is the case because consumers who produce their own electricity do not buy it from their utility. Fewer kilowatt-hours sold means lower revenues. In order to compensate for reduced income, utilities argue that they will be forced to raise rates on their remaining customers, thereby driving more people to use DG systems.[36] According to this worst-case scenario, the culmination will be a "death spiral" for the utilities, which will be stuck with grid maintenance costs, fewer customers, reduced demand, and a whole host of stranded assets.

For a utility, the loss of revenue from traditional sales does not necessarily correspond with proportionally lower costs to supply electricity, which is why rate increases seem virtually inevitable in the face of growing solar usage. Only one portion of a consumer's cost of a kilowatt-hour is the cost of the electricity itself. This cost is "bundled" with all the other costs associated with the ongoing maintenance of the grid and all of its services, such as providing transmission and distribution equipment and services, maintaining backup generation capacity, reselling the customer's excess energy, balancing supply and demand in subsecond intervals, and providing voltage and frequency control.[37] These services may be provided by a single utility or multiple power-producing and marketing entities. Either way, their costs are reflected in the bundled price. Homeowners and firms that have a close or exact balance between their net usage and excess generation, or that buy power from utilities only at peak times or during emergencies, are seen by the utilities as receiving all these benefits without fully paying for them. In fact, in California the state government does not allow new or additional demand charges, standby charges, customer charges, minimum monthly charges, interconnection charges, or other charges for customers who have installed solar PV or other on-site generation systems as an inducement for greater use of solar energy.[38]

When "net metering" is added to the equation, the financial issue is further highlighted. Net metering allows customers to sell excess electricity back to the utility, while receiving electricity when their own PV systems are not producing power sufficient to meet demand. The idea is to base the size of an on-site system (and thus the cost of such a system) on total annual usage, not peak usage. This practice allows the incoming and outgoing kilowatt-hours to be balanced over time due to seasonal variation in both the demand for electricity and the amount of sunlight available (while solar resources are highly variable at any given time, they are relatively predictable on an annual basis). In places where consumers are paid the full bundled rate by the utility for any excess electricity supplied into the grid (this increases the value for those using DG), it effectively means that utilities are paying businesses and homeowners for costs that they do not incur. Even in cases where utilities pay less than the full bundled rate, "utilities project that under current policies, solar users' savings add about $1.3 billion to non-solar users' bills. In other words, people who don't want or can't afford to install solar are paying for those who do."[39]

The conclusions that utilities have reached, as a result of needing to operate with these solar energy use incentives in effect such as net metering, limitations on cost recovery (e.g., standby rates, demand charges), and regulatory mandates are that: 1) the balancing and backup services that utilities provide to DG customers are needed and valuable; 2) it's not feasible for customers to provide these services for themselves; 3) it is unfair for DG customers to avoid paying for these services, shifting the cost to non-DG customers; and, 4) DG customers should be paying their fair share of the cost of the grid services that their utilities provide.[40] Without a fair reallocation of costs, argue utilities, the "death spiral" may eventually overwhelm them.

The counterargument to the utility point of view suggests a very different understanding of distributed generation and utility business models. One of the strongest arguments is made by Amory Lovins of the Rocky Mountain Institute. He suggests that the appropriate response to utilities' financial troubles is not lamentation, but rather celebration. Pointing out that disruptive technologies are expected to upend long established practices, it is not surprising that "utility companies that refuse to let go of an archaic system are losing money" at a time when "twenty-first-century technology and speed collide with twentieth- and nineteenth-century institutions, rules and cultures." As such, in his view there is no good reason to shield utilities from facing state-subsidized competition from solar and other renewables.[41]

In addition, Lovins suggests that concerns about grid-stability and imbalances are often overblown for the sake of promoting favorable consideration from state regulators:

> Well-stoked fears of grid instability and unreliability due to renewable power are as widespread as evidence for them is unfindable. In the Central European grid, where pervasive electricity trading helps operators choreograph the ever-shifting mix of renewable and nonrenewable supplies, German electricity (23 percent renewable in 2012) and Denmark (41 percent) are the most reliable in Europe—about ten times better than in the United States (whose 2012 electricity was 6.6 percent hydro and 5.3 percent other renewables). Even on the edge of the European grid, Spain (48 percent in the first half of 2013) and Portugal (70 percent) kept their lights on just fine.[42]

This argument can be extended further, as DG customers provide a service to the utilities that the industry fails to acknowledge—namely, the supply of electricity that would otherwise have to be procured by the utility to meet customer demand. Every kilowatt-hour provided by on-site solar is a kilowatt-hour that a utility does not have to purchase or generate itself. Since solar PV generates its highest power output during times of peak usage, DG users are allowing utilities to avoid buying relatively high-priced electricity in wholesale spot markets. In other words, the ever-growing demand for electricity in the United States is in part being met not by the customary construction of new power plants, but by a new class of consumer/producers, or "prosumers." These avoided utility costs represent an additional component of the cost/benefit analysis that informs the debate over the relative value that utilities and consumers provide to each other.

The potential responses to this evolving situation of a changing landscape of electrical power producers involve changes by both market actors and policy-makers. Utilities are responding to these ongoing and coming changes in several ways. One approach is to take actions to cut costs. In Europe, where DG is more widespread than in the United States, large utilities such as *RWE* and *E.on* have already begun to do this. *E.on*, for example, is implementing a series of cost-cutting measures that are designed to save about 1.5 billion Euros, and this will include cutting 11,000 full-time equivalent jobs.[43] As one US industry representative noted, "The scope to take 10–20% out of the cost base of companies in the sector is definitely there. It would provide some room for the longer-term sustainability of companies as they adjust their strategies ... accelerating the pace of development in things like geospatial technology, mobility tools, smart grids and sophisticated scheduling and warehousing can all provide a springboard for major cost savings."[44]

Another response by utilities is to move away from selling only electricity and begin to also sell electric power services. This product diversification can include activities tied to energy efficiency, such as conducting energy audits; installing and/or financing DG systems; making energy efficiency upgrades; and assessing operations and maintenance for potential cost reductions. Such services can be financed by fee-for-service or performance-based contracts, where payment is based on energy savings achieved. A more far-reaching approach, dependent upon further development of smart grid technologies, is for utilities to move away from selling kilowatt-hours and move toward the sale of outcome services such as heat, light, and screen time.

Utilities might also seek to become the owners of distributed generation assets. This type of an arrangement, essentially a power purchase agreement, has been widely adopted in the purchase of large PV systems for commercial and industrial customers, but it has been undertaken largely without utility participation. Instead, "third-party" financiers facilitate sales of solar PV by serving as owners of the generation systems and selling the electricity produced to consumers. This approach is also growing in the residential market, with companies such as Solar City serving as the third-party electricity broker. This approach has allowed for the rapid growth of solar PV in both the commercial and residential sectors, as it allows energy users to avoid not only significant upfront technology installation costs, but also the responsibilities and risks associated with being the owners and operators of electricity generation systems. Utilities have generally not pursued this approach to DG. Becoming the owners of numerous on-site generation systems has not historically been part of utilities' business models, so it may be expected that this is not widely considered to be part of their future. Another element to consider is that deregulation in some cases prevents this type of utility role. In states where substantial retail deregulation has occurred, utilities are usually prohibited from owning power generation. The idea behind such restructuring, which has also been supported at the federal level by FERC, has been to sever the ties between generation and transmission/distribution in order to increase competition in the marketplace for power generation. In dealing with a business model characterized by utilities owning big, central power plants, such a measure has been effective in opening up the market to greater competition. However, this may not make as much sense in an environment where distributed solar power is relatively widespread.

As these examples suggest, the realm of energy policy is inherently connected to the way in which business models are developed and maintained. To the extent that utilities and power grid managers are going to implement changes to deal with DG, they will have to engage with regulators and policymakers to effect these changes. One such area to be visited may involve ownership of generation assets. Other newly emerging solutions involve reallocation of costs. (A larger issue, but one not yet part of the policy debate, involves the extent to which utilities should continue to be considered natural monopolies, and strictly regulated with regard to electricity rates, as consumers increasingly come to generate their own electricity.) A 2013 survey of utility representatives, along with an Edison Foundation report issued in the same year, found that one of the top solutions recommended (by utilities) is changes to tariff structures that will allow utilities to charge DG customers for grid services they receive but for which they do not fully pay.[45] These charges could come in the form of what are commonly called backup charges or standby rates (another emerging term is "network usage charge"), covering needs during planned or unplanned outages, needs that exceed supply from DG systems, and delivery of such service. Demand charges, on the other hand, are assessed based upon customer maximum demand during a given period. These charges allow for utilities to pay for generation capacity that is needed to supply electricity during the times of peak demand. Commercial and industrial users tend to pay demand charges, while residential customers do not, but the growing use of various DG technologies in all sectors is leading electricity suppliers to suggest that demand charges (as well as standby rates) need to be more broadly assessed in order to pay for maintaining and operating the nation's power grid systems.

The downside of standby rates and demand charges is that while they provide an economic benefit to utilities for providing electricity services, they also provide a disincentive to use on-site generation. Such charges will reduce the use of solar energy and will diminish the economic, social and environmental benefits to be gained—providing a hedge against rising fuel costs and peak demand charges, creating new jobs in the solar industry, and reducing air pollution and carbon emissions. Since the possible new charge imposition or increases of such existing charges represent a "zero-sum game," they are particularly contentious public regulatory policy and political issues.

At both the federal and state levels, there are political challenges to policies that support renewables, including sunsetting the ITC and PTC, repealing state RPS mandates, cutting back net metering, and adding backup or standby charges to the utility bills of solar PV customers. This effort to roll back support for solar and wind power has been somewhat successful for the federal PTC, which expired in 2014 and was then extended for only a few more years. It is argued that the wind and solar industries do not need their federal incentives any longer, due to their substantial market growth and greatly reduced technology costs. Elimination of the tax incentives created to jumpstart the industry would almost certainly cause the adoption rate of wind and solar to slow down to some extent, but it is further argued that this will diminish the pressure on utilities and grid managers to address the technical and financial challenges resulting from variable, distributed generation. Moreover, allowing solar and wind to compete in the market without federal support would, according to this argument, create greater market efficiency instead of steering investment toward sectors that gain added financial benefits only as a result of tax law advantages. For example, the PTC pays for kilowatt-hours whether or not they are needed in a given area at a given time.

At the same time, supporters of renewable energy point out the numerous environmental, economic and social benefits of using renewable energy, as well as the long-established tax breaks and subsidies provided—on a permanent basis—to the oil and gas industries. Likewise, in several states—North Carolina, Kansas, Arizona, Ohio, Washington and others—changes to net metering laws, RPS mandates, and backup charges have been sought by organizations such as the American Legislative Exchange Council (ALEC, a membership group of conservative state legislators who develop and share legislative proposals), along with utilities, and the Koch brothers (David and Charles Koch, the billionaire industrialists and political activists, who provide funding to conservative candidates and organizations and support efforts to stop or roll back legislation on renewables and climate change). In Arizona, state regulators approved a $5/month charge to rooftop solar users in 2013. This fee imposition was a clear victory for the power utilities because it acknowledged the principle that DG customers have to pay for services they receive. It was a victory for prosumers as well, inasmuch as the fee was far less than the $50/month sought by Arizona Public Service, the state's largest utility.[46]

In this political climate, the renewal and/or extension of the PTC and ITC is uncertain, as are the multiple challenges to various state laws incentivizing renewables. Because the energy policy environment is in such flux, it remains to be seen if the evolution of solar and wind adoption will be led primarily by public policy measures or by market developments in the absence of policy support. Most likely a combination of the two influences will operate in an ongoing ebb and flow.

The Promise (Largely Unfulfilled) of Biofuels

The search for alternative liquid fuels in the United States has been based on everything from national security threats and concerns with reliability, to market volatility and environmental values. Ever since the oil embargo imposed by OPEC in 1973, the price and reliability of oil has been subject to events that are often beyond the control of the United States. American policy toward the oil-producing states in the Middle East has since this time largely revolved around the subject of ensuring oil supplies. Iraq's invasion of Kuwait in 1990 ushered in an era of extensive American military involvement in the region, which was driven in large part by energy insecurity. Speaking about the national security costs of reliance on oil, based on the country's experiences in Iraq and Afghanistan, Secretary of the Navy Ray Mabus said in 2011, "There are great strategic reasons for moving away from fossil fuels." This is because, "We buy our energy from people who may not be our friends. We would never let the countries that we buy energy from build our ships or our aircraft or our ground vehicles, but we give them a say on whether those ships sail, whether those aircraft fly, whether those ground vehicles operate because we buy their energy." Moreover, he added, "It's costly. Every time the cost of a barrel of oil goes up a dollar, it costs the United States Navy $31 million in extra fuel costs. When the Libya crisis began and the price of oil went up, the Navy faced a fuel bill increase of over $1.5 billion."[47] Diversifying the resources needed for transportation—in terms of both fuel source and geographic source—would ideally diminish the numerous liabilities associated with oil dependency. And in practical terms, this has largely meant pursuing the development of domestically-produced ethanol since the 1970s to enhance the nation's energy security.

The production and use of ethanol in the United States has increased greatly. In 2000 only 1.6 billion gallons of ethanol were produced. By 2013, that level had risen to 13 billion gallons (almost all of this was used domestically), making the United States the world's largest producer of ethanol.[48] In 2011, there were 93 million acres used to grow corn in the US, and roughly 40% of this went into producing ethanol. (It should be noted that cattle feed is a by-product of ethanol production, so the net acreage going only to fuel is less than 40%.) In that same year, 209 biorefineries were in operation processing the corn into fuel.[49]

The major driver behind the growth of ethanol production and use in the United States can be found in the realm of public policy. Congress first enacted subsidies to producers of ethanol in 1978 with the *Energy Tax Act*, which provided a tax exemption of $.40/gallon. From this time through 2011, when the subsidy was eliminated, producers of ethanol received anywhere from 40 to 60 cents a gallon, as subsequent legislation (the *Alternative Motor Fuels Act*, the *Energy Policy Act of 1992*, the *Energy Policy Act of 2005*, among others) both extended the subsidy and revised the amount of the tax credit.[50] In 2011, the tax credit was $.45/gallon for corn ethanol (and up to $1.01 for cellulosic ethanol, for which the subsidy remains). A 2010 study by the Congressional Budget Office found that biofuel tax credits cost $6 billion a year in foregone revenue, which meant that taxpayers paid $1.78 per gallon of reduced use of conventional gasoline.[51] In addition to providing a tax incentive, Congress also protected the domestic market by placing a $.54/gallon tariff on ethanol imported from other countries. This measure, passed in 1980 and eliminated in 2011, was largely designed to curb imports of sugarcane ethanol produced in Brazil.[52] Apart from support for ethanol production, Congress also provides several billion dollars per year in support to growers of corn, the most widely grown crop in the country—ranging from $2–10 billion a year between 1995 and 2012.[53]

Congressional elimination of tax and tariff benefits targeted toward a particular industry is a rare event in American politics, but in this case the industry did not need to fight for these measures because other provisions in US law ensured a market that would be expected to grow. What occurred was a switch of policy tools, in which incentives and protection were replaced by mandates. These mandates have required: 1) using oxygenate additives to reduce air pollution; and, 2) increasing the production and use of renewable biofuels.[54]

The use of oxygenate additives in gasoline is mandated by the *Clean Air Act*. This is the case because "reformulated" fuel that contains an oxygen additive will burn more efficiently, releasing fewer pollutants such as carbon monoxide into the air. The *Clean Air Act Amendments of 1990* introduced the mandate for oxygenated additives in the US gasoline supply. Metropolitan areas that exceed a particular level of carbon monoxide in the air—mostly in California and on the East Coast, along with big cities like Houston and Chicago—are required to use reformulated gasoline. While the additive known as MTBE initially appeared that it would emerge as the most widely used oxygenate to meet this requirement, problems involving leaking storage tanks adversely impacted water quality in many places, leading to its diminished use or outright ban throughout the country. In its place, ethanol came to be the preferred additive used in the domestic gasoline supply.

While the oxygenate requirements greatly expanded the use of ethanol in the United States, it is the Renewable Fuel Standard (RFS) that produced a sea change with respect to the production and use of ethanol, prompting a massive expansion of the industry.[55] The RFS was established by the *Energy Policy Act of 2005*, and then the targets were ramped up in the *Energy Independence and Security Act of 2007 (EISA)*. The law requires that biofuels be mixed into the supply of gasoline, and that the total level used continually increases. Nine billion gallons of ethanol were blended into the gasoline supply in 2008, and 14 billion gallons were used 2011. As shown in Table 5.1, the law calls for 36 billion gallons of ethanol and advanced biofuels to be part of the fuel supply by 2022.[56]

The policy can boast a degree of success, as the RFS targets were met every year through 2011, displacing an increasing amount of conventionally produced gasoline. In 2012, the amount of ethanol blended into the domestic liquid fuel supply fell short of the 15.2 billion gallon goal by 1.9 billion gallons, and it has continued to fall short of the legislatively-mandated goal ever since this time.[57] The unavoidable tradeoffs associated with ethanol production, however, have diminished the early enthusiasm for the use of purpose-grown crop conversion into ethanol as the primary source. The collateral costs of rising ethanol use have been documented in a growing chorus of criticism amidst the rapid expansion of biofuel production. First of all, as a political matter, it is clear that successful lobbying provides taxpayer support to grow an industry whose size would be far smaller than it currently is. Second, a body of scientific literature has emerged demonstrating that the energy savings and environmental benefits originally thought to result from ethanol use are not as large as originally believed. In fact, when a life-cycle perspective is applied to the cost-benefit assessment of fuel sources, it is possible that in some cases the production and use of ethanol can result in net energy and environmental setbacks.[58] Lastly, the increased use of land to produce fuel instead of food has driven up the price of corn, and this has subsequently caused an increase in prices of other food sources.[59]

TABLE 5.1 RFS Mandate (in gallons)

Year	Ethanol	Advanced	Total
2008	9.00	0.00	9.00
2009	10.50	0.60	11.10
2010	12.00	0.95	12.95
2011	12.60	1.35	13.95
2012	13.20	2.00	15.20
2013	13.80	2.75	16.55
2014	14.40	3.75	18.15
2015	15.00	5.50	20.50
2016	15.00	7.25	22.25
2017	15.00	9.00	24.00
2018	15.00	11.00	26.00
2019	15.00	13.00	28.00
2020	15.00	15.00	30.00
2021	15.00	18.00	33.00
2022	15.00	21.00	36.00

Source: Renewable Fuels Association.

These critiques of biofuel production speak to the varied costs (some unintended and/or unexpected) of a public policy goal aimed at greater energy security. These costs (beyond those involving taxpayer support to ethanol producers and corn growers) tend to revolve around three major areas: 1) the energy inputs required to produce ethanol and the energy outputs ethanol provides; 2) the carbon dioxide reductions available from using ethanol as opposed to gasoline; and, 3) the increased use of land to produce fuel instead of food.

Energy Inputs and Outputs

One gallon of corn-based ethanol contains about two-thirds of the energy content of gasoline, so in order to do the same amount of work as gasoline, 1.5 gallons of ethanol are required. However, what must also be considered is the amount of energy used to produce each gallon of ethanol. In other words, how much energy must be expended in order to produce or acquire more energy? This concept is known as the "energy return on energy invested," or EROEI. This ratio is usually expressed as a number. For example, an EROEI of 10 means that for every unit of energy expended, there is a return of 10 units of energy. If the EROEI is above a level of 1 for a given resource, then it would represent a net energy yield. If the EROEI falls below a level of 1, the resource would involve a net energy loss.

Ethanol's EROEI can vary greatly, as it depends upon several factors. These include the fuel required to plant and harvest the corn, to acquire and apply fertilizer, to process the corn into fuel in biorefineries, and to transport the fuel to the next point in the supply chain. Because these energy needs can be large, the EROEI of ethanol is not definitively considered to be positive. For example, some analyses indicate that ethanol has an EROEI below a level of one, meaning that ethanol does not even replace the energy content needed to produce and deliver it.[60] Several other studies, by contrast, have found that ethanol has a positive EROEI. Not surprisingly, the Renewable Fuels Association, the industry's member association, argues that the EROEI of ethanol most certainly provides a net gain, approaching a level close to two.[61] A meta-analysis published in 2011 looked at previous studies in an effort to look at the range of findings in the scientific literature. This study found that the net energy from ethanol ranged from 0.64 to 1.18, depending upon the fuels and processes used to produce the ethanol.[62] By way of contrast, the EROEI for oil, which has declined greatly over the decades, from an estimated high of 100 in the 1930s in the United States, currently ranges from roughly 2 to 10, depending upon factors such as the techniques used to produce the oil and the length of the supply chain.[63] Therefore, according to the meta-analysis, ethanol offers a very small return in comparison to petroleum, and a negative return in some cases. Moreover, the analysis stated that, in considering the actual production processes employed in the country's 127 biorefineries in 2009, only about 5% of total production could be considered net energy efficient. The conclusion was that even in the best scenario, the country is "gaining an insignificant amount of net energy" from ethanol.[64]

Greenhouse Gas Emissions

Similar to the net energy concern, the performance of ethanol with regard to greenhouse gas (GHG) emissions may be less than promised and expected. In theory, biofuels can be understood as carbon neutral. The idea is that even though carbon dioxide is emitted when they are burned, the corn or other materials used to produce the ethanol originally absorbed that carbon dioxide from the atmosphere. Therefore, no new net emissions occur. However, in practice this argument may not stand up, as it fails to take into account the full life cycle in the production and use of ethanol. Additional carbon dioxide emissions are created by growing corn, converting it into fuel, and transporting the fuel to the point of use. Such calculations may also be impacted by considering the change in land use from some other activity to corn production. For all these reasons, the carbon dioxide balance sheet does not always emerge as favorably as hoped or anticipated.

The argument over the proper accounting of carbon emissions is robust and ongoing. The scientific literature finds in general that ethanol use results in fewer emissions than gasoline, but much depends upon how the fuel is produced and how far it is transported. A study conducted by the Argonne National Laboratory found that if coal is used to provide power to a biorefinery, then the GHG impact from using ethanol actually represents a 3% increase over gasoline. On the other hand, using wood chips to power the same plant would provide a 52% reduction in emissions.[65] A study published in *Science* found that ethanol use resulted in a 13% reduction in emissions compared to using gasoline, but the authors also stated that this would vary from place to place.[66] The Renewable Fuels Association has argued that these studies miss a key point, which would result in finding much larger GHG reductions. The organization points out that, "fossil fuel producers are going farther and deeper—and using more energy and emitting more GHGs—to extract new sources of oil. The Association argues that these are the marginal sources of oil against which ethanol should be compared, since both unconventional oil and ethanol are the newest entrants into the fuel pool."[67] Incorporating an approach such as this in estimating the GHG impact of ethanol would most certainly lead to a result that finds a more significant reduction. It would also suggest that the EROEI is higher than the academic literature has determined.

Food or Fuel?

One of the most significant issues involving ethanol that has emerged in recent years has been the impact on land use and the production of food. As a large proportion of the corn grown in the United States (and European Union) is destined for ethanol production, a vigorous debate has emerged on both the societal impacts and the ultimate wisdom of increasing the use of arable land to produce fuel instead of growing food.

Though all the land used for ethanol production does not necessarily come at the expense of food production, the rising price of corn and other staples, which have seen periods of sharp price spikes in recent years, suggests that food production is suffering to some extent. In 2005, before the RFS mandate went into effect, the average price of a bushel of corn sold in the United States was $1.96. In August of 2012 a market high was reached with a monthly average of $7.63 a bushel. The price since declined to $3.79 a bushel in 2015.[68] These price spikes appear to have had an inflationary impact on the price of food in general, in the United States and around the world, according to several well-researched studies.[69] With less corn available for use in food products, consumers appear to have turned to other substitutable commodities such as wheat, rice and soybeans, which helped cause the price of these goods to more than double between 2005 and 2013.[70]

The growing use of land to produce fuel and the subsequent market volatility appear to have caused further unintended consequences around the world, namely, food riots and social unrest in many places in the developing world. Egypt, Cameroon, Mozambique, Senegal, Haiti, India, and Bangladesh have faced rioting in the past several years attributed to dramatic food price spikes. Even the protests that resulted in the Arab Spring in 2011 may have been triggered in part by a response to the high price of food.[71] An analysis done in 2011 looked at riots worldwide, whatever the cause, and found that they have been widely correlated with rising food prices.[72] Since food is a necessity while gasoline is much more of a convenience or even a luxury good, the situation that has evolved is one in which those who are financially better off seem to be benefitting at a cost to those in poverty, especially in the developing world where people spend a larger proportion of their income on food than is the case in the more highly developed regions.[73] "For the 1.2 billion people who make $1.25 or less a day and spend 50% to 80% of their income on food, price rises mean hunger and less money for education and health care."[74] As one food security analyst put it, "It may be the case, therefore, that the poor go hungry so the wealthy can drive bigger cars farther ... Without adequate safeguards, further expansion of biofuels will mean an unpalatable trade-off between cars for the rich and starvation for the poor."[75]

The Politics of Ethanol

The ethanol industry in the United States boasted a total value to the US Gross Domestic Product in 2011 of $43 billion.[76] The stakes are clearly high, and because the industry is driven to a great extent by public policy, the politics of ethanol are particularly intense. The farm lobby in the United States has been particularly successful in this regard, winning billions of dollars in federal support for producing corn, the most widely grown crop in the United States. This support includes direct payment to farmers, price supports for crops produced, and crop insurance. A Congressional report in 2010 found that direct payments totaled $5 billion per year, with other supports at roughly $1 billion a year.[77] One of the features of this policy is that direct payments are based on acreage planted, and the level of the payment does not take into account either yields or market prices. This has prompted two critiques. First of all, the largest benefits go to the largest farms, many of which are owned by commercial firms, not family farmers. What this means is that the most economically vulnerable group of farmers tend to receive the least support. The Environmental Working Group, which has been a vocal critic of US farm subsidies, says that "the top 10 percent of farmers collected 74 percent of all subsidies between 1995 and 2010 (not only direct payments), amounting to nearly $166 billion."[78] A second critique is that payments are made without regard to crop yields or prices. Therefore, the federal government may be providing financial support at times when its necessity is questionable. Consider economic conditions since 2005: corn and other food prices have been relatively high, the use of ethanol is mandated by federal law, growing corn has been profitable, and the United States has a sizable budget deficit. In this context, the critics have a valid point, asking if it makes sense continuing to subsidize the nation's corn growers so generously.

The politics of ethanol are also significantly affected by the US presidential nomination process. Long before a single ballot is cast in the primaries, the Iowa caucuses—the first such contest in securing the nomination—command the attention of presidential aspirants who spend months prior to the caucuses organizing and giving speeches in the state. Success in Iowa is considered crucial for a candidate to remain viable and stay in the presidential race. For this reason, candidates from both major political parties consistently spend a great deal of time in Iowa, and routinely voice support for ethanol when they campaign (Iowa produced 2.2 billion bushels of corn on 13 million acres of land in 2013).[79] Hillary Clinton, Jeb Bush, Scott Walker and others running in 2016 have been supportive of the Renewable Fuel Standard.[80] Before them, Barack Obama and Mitt Romney were both supportive of the mandate in 2012. These are only the most recent examples of what has been a long succession of presidential candidates offering support for the nation's ethanol policy. The rare exception has been John McCain, an opponent of ethanol production and farm subsidies in general. He has been the only major candidate to publicly challenge the subsidies and mandates that boost ethanol usage, arguing that it does not improve either energy security or air quality, only the chances of securing votes. In 2000, McCain said that "ethanol subsidies should be phased out, and everybody here on this stage, if it wasn't for the fact that Iowa is the first caucus state, would share my view that we don't need ethanol subsidies. It doesn't help anybody."[81] McCain changed his position somewhat in 2008, during his second presidential campaign, when he came out in support of ethanol production as a solution to energy security and GHG emissions, though he still remained opposed to subsidies. Surprisingly, prior to the election in 2012, polls conducted in Iowa found a decline in the number of voters who considered support for ethanol to be a major issue in their voting.[82] Perhaps unflinching support for ethanol in Iowa is now a bit less important to the state's voters, though judging from the crop of presidential candidates trekking to Iowa in 2015 and 2016, and their outspoken support for ethanol mandates, it appears that candidates are not taking any chances.

The political coalitions that have emerged in response to ethanol production and mandated use have been unlike others in the energy sector. The debate in the United States between supporters of renewables (particularly wind and solar) and fossil fuels (especially oil and gas) looms large in the United States, splitting along typically expected lines of right (fossil energy) vs. left (renewables). Supporters of each see their favored resources as driving an American energy renaissance that could lead to the elusive goal of energy security, while seeing little merit in the aims of the other side. In the specific case of ethanol, however, the political divide is somewhat different, pitting environmentalists, domestic oil producers, ranchers, red-state governors, and advocates for the poor against grain farmers, biorefiners, OPEC-haters, energy independence advocates, other groups of environmentalists, and until late 2013, the EPA. The economic interests and political aims of these various groups and interests have created a situation in which the first of these coalitions seeks to eliminate or reduce the fuel blending requirements in the Renewable Fuel Standard, while the second coalition wants to maintain them.

In 2012, the political landscape regarding ethanol shifted in response to economic and political pressures, in particular the rising price of corn, and the lack of a similar rise in gasoline usage. As the cost of feed corn continued to increase (a severe drought in 2012 limited corn production that year), livestock and poultry producers saw their costs jump, leading to higher consumer costs. These circumstances prompted the governors of several states (Texas, North Carolina, Georgia, Virginia, New Mexico, Arkansas, Maryland, Delaware, Utah and Wyoming), to request that the EPA provide a temporary waiver of the Renewable Fuel Standard.[83] The EPA can waive the RFS if meeting the targets causes either severe economic or environmental harm, but the agency denied the request, stating that, "this year's drought has created significant hardships in many sectors of the economy, however, the agency's extensive analysis makes clear that Congressional requirements for a waiver have not been met and that waiving the RFS would have little, if any, impact on ethanol demand over the time period analyzed."[84]

The EPA decided to revisit its position the following year after biofuel usage fell short of the 2012 goal by 1.9 billion gallons.[85] It proposed reducing the RPS levels for 2014 from 18.1 billion gallons to 15.2 billion gallons. The problem has been that high gasoline prices, more fuel-efficient cars, and a somewhat sluggish economy translated into less gasoline use than expected. With total gasoline usage rising more slowly than the mandated level of ethanol blend, there was a concern that the mandate could not be met. Adding to the problem is a technological pressure, known as the "blend wall," which is the level at which ethanol can be added to gasoline without causing harmful effects on vehicle engines. The petroleum and automobile industries have argued that while newer cars can handle a blend of greater than 10% ethanol (such as E15 or higher), most older cars and trucks cannot burn that fuel without potential harm to engine parts. EPA also acknowledges the issue of a blend wall in its mandates. However, this potential problem is not without controversy. The Renewable Fuels Association, for example, argues that scare campaigns, "junk science," and strong-arm tactics by the oil industry to prevent gas stations from offering fuel with higher blends of ethanol have limited the availability of E15 and E85, which could power millions of cars without a problem.[86]

An additional problem befalling the RFS is the mandate to use advanced biofuels and cellulosic ethanol. The RFS calls for the use of corn ethanol to be capped at 15 billion gallons in 2015. After that, the increase in biofuel use would have to come from advanced biofuels and cellulosic ethanol. Cellulosic ethanol, for example, is considered to offer a great improvement over corn ethanol. The feedstock is abundant and can come from agricultural and forest waste products. In addition, the EROEI and greenhouse gas emissions associated with cellulosic ethanol are far better than those for corn ethanol. A National Academy of Sciences study found that switchgrass offers an EROEI ratio of 5 to 1, while reducing CO_2 emissions by 94% compared to conventional gasoline.[87] The production of these advanced biofuels, however, has not materialized as quickly as anticipated. The industry has not been able to produce as much as Congress hoped for and envisioned back in 2007. These liquid fuel sources are not yet cost-effective or scalable to the necessary levels. To that end, the EPA has been making changes to revise its mandated targets. The cellulosic ethanol target for 2013 was revised downward from the originally mandated 1 billion gallons to 6 million gallons, and then again in 2014 to what was actually produced in 2013—only 810,000 gallons.[88]

It is expected that the targets for conventional ethanol, along with advanced biofuels and cellulosic ethanol, will continue to be reduced as necessitated for the foreseeable future. In this regard, the EPA is simply responding to the technical and economic realities that Congress did not and could not anticipate accurately. Without these revisions, as the mandated level of biofuel use has become unattainable in the required timeframe, the EPA would have to find firms in widespread non-compliance with the law through no fault of their own. The biofuel goals would have become virtually meaningless and unenforceable.

The Times They Are A-changin' (Partly)

In considering the possibility of broad transformation in the energy sector resulting from extensive growth in the use of renewable energy sources, it is clear that rather rapid change is indeed occurring, though it is quite unclear just how transformational the changes will prove to be on energy markets and consumers in the long run. The market for liquid fuels will continue to remain dependent upon gasoline for a very long time into the future. Technology options that are feasible and have a hope of being commercially competitive in price are limited in number, while a strong and well-established market built up around oil and gasoline exists and continues to serve the needs of a great many people. It is fair to say as well that the public policy measures enacted designed to promote alternative fuels and diminish the primacy of oil have run up against a variety of technical, political, and market obstacles. Moreover, the policy environment involving oil remains geared toward the continual attainment of greater production capacity. The transportation sector still awaits (either eagerly or with concern, depending upon to whom you pose the question) its disruptive technology, its own version of wind and solar power.

By contrast, the growing opportunities (and threats) in the electric power market suggest that the restructuring of the electric power market is very likely, if not inevitable. Whether it will be ultimately attributable to technological change, global climate change, policy change, consumer behavior change, or some combination of all of these, the evolution of the electric power system is clearly underway. This development is unsurprising, considering that the electricity sector still uses the same basic model that Thomas Edison developed at the dawn of the electrification phenomenon. Think about it for a moment; if Alexander Graham Bell came back today and saw our communications infrastructure, he would not recognize it. The smartphone offers capabilities for talking, text, internet, email, photos, video, music, games, and thousands of other applications, all available without being connected to a wire. This would amaze him. Thomas Edison, on the other hand, would be able to recognize the electric power grid in a second, and could arguably say, "This looks just like what I built, only bigger and somewhat more sophisticated." Just as the telecommunications sector underwent massive change due to technology development, market innovation and policy measures adopted to increase competition, so too is the electric utility industry moving through a similar period, one which promises a great deal of turbulence and upheaval.

A final point needs to be made here. In considering the obstacles presented by the expanded use of renewable resources such as biofuels, solar and wind it is important to note that this does not imply opposition to their growing use, or a suggestion that renewables represent an unfruitful path to pursue. While there are barriers in the efforts to produce and use greater levels of wind and solar power, and to increase the use of advanced biofuels, there is little doubt that many or most of these barriers will be overcome in time (though new challenges are almost certain to emerge). Technologies, business models, markets, and public policy, along with consumer preferences, all continue to develop and mature in the energy area. Moreover, even though some of the costs associated with their use may be high compared to alternatives, these costs may prove to market participants, policymakers and citizens/consumers to be well worth it. The benefits of obtaining additional domestically-supplied energy from cleaner sources may provide enough economic, environmental and/or national security benefits to justify prices and likely tradeoffs to be made along the way. After all, every energy source has its costs, as none provides the complete answer to the achievement of energy security.

Notes

1 Amory Lovins, "Let's Celebrate, Not Lament, Renewables' Disruption of Electric Utilities," *RMI Outlet*, February 6, 2014.
2 American Wind Energy Association, *U.S. Wind Industry Annual Market Report 2014*; and US Energy Information Administration (EIA), *Annual Energy Outlook 2015*.
3 Solar Energy Industries Association (SEIA), *Solar Market Insight Report 2015 Q2*; and *Solar Market Insight Report 2011 Year in Review*.
4 The Regulatory Assistance Project, *Electricity Regulation in the U.S.: A Guide*, March 2011.
5 Joshua Green, "The Elusive Green Economy," *The Atlantic*, July/August 2009.
6 CohnReznick Think Energy, "Case Studies, Toyota," www.reznickthinkenergy.com/case-studies (accessed June 10, 2014); and Solar Energy Industries Association (SEIA), "2014 Top 10 Solar States," www.seia.org/research-resources/2014-top-10-solar-states (accessed September 15, 2015).
7 SEIA, "New Mexico Solar," www.seia.org/state-solar-policy/new-mexico (accessed September 15, 2015).
8 Wind Power Monthly, "The 10 Biggest Turbines in the World," www.windpowermonthly.com/10-biggest-turbines (accessed June 10, 2014).
9 Go Solar California, "California Solar Initiative Rebates," www.gosolarcalifornia.ca.gov/csi/rebates.php (accessed June 10, 2014).
10 Energy Trust of Oregon, "Incentives and Financing for Solar," http://energytrust.org/renewable-energy/ (accessed September 15, 2015).
11 SEIA, *U.S. Solar Market Insight Report 2014*. In Oregon, the 104 kW "Solar Highway" project on I-5 just outside of Portland demonstrates the limitations of big solar PV investments in certain places. At 10:00 AM on Christmas Day of 2013 the project was producing only 4.8 kilowatts of electricity, less than 5% of its potential. The system's peak in the preceding week was 6.3 kW of power, and its total generation over that same week was 480 kilowatt-hours, enough to power only five 100-watt light bulbs for one hour. This level of generation represents the low end of yearly averages, but it is not uncommon in the winter months. See: http://live.deckmonitoring.com/?id=solarhighway (accessed June 10, 2014).
12 EIA, *Electric Power Monthly*, March 2014; Massachusetts Office of Energy and Environmental Affairs, "Solar Carve Out, SREC I," www.mass.gov/eea/energy-utilities-clean-tech/renewable-energy/solar/rps-solar-carve-out/ (accessed October 11, 2015); and SREC Trade, "Massachusetts Market Prices," www.srectrade.com/srec_markets/massachusetts (accessed October 11, 2015).
13 Texas State Energy Conservation Office, "Texas RPS," www.seco.cpa.state.tx.us/re/rps-portfolio.php (accessed June 10, 2014); US Department of Energy, "Energy Efficiency and Renewable Energy, Installed Wind Capacity," www.windpoweringamerica.gov/wind_installed_capacity.asp (accessed June 10, 2014).
14 Texas Public Utilities Commission, "Competitive Renewable Energy Zones," www.texascrezprojects.com/overview.aspx (accessed June 10, 2014).

15 Bonneville Power Administration, "BPA Balancing Authority Load and Total Wind, Hydro and Thermal Generation," September 10–17, 2015, http://transmission.bpa. gov/business/operations/wind/baltwg.aspx (accessed September 17, 2015).

16 "BPA's Wind Power Cutoff Sends a Troubling Signal," *The Oregonian*, May 19, 2011.

17 William Pentland, "Breaking Wind Records in Texas," Fierce Energy, www.fierceenergy. com/story/breaking-wind-records-texas/2015-09-15 (accessed September 15, 2015); and "Wind Energy Causes Negative Electricity Prices," *Search for Energy*, www.searchfor energy.com/blog/wind-power-causes-negative-electricity-prices/ (accessed September 15, 2015).

18 "How to Lose Half a Trillion Euros," *The Economist*, October 12, 2013.

19 Christopher Joyce, "Power Grid Must Adapt to Handle Renewable Energy," *NPR Morning Edition*, March 12, 2012, www.npr.org/2012/03/12/148318905/renewa ble-energy-throws-power-grid-off-balance (accessed June 10, 2014).

20 Cory Budischak, DeAnna Sewell, Heather Thomson, et al., "Cost-minimized Combinations of Wind Power, Solar Power and Electrochemical Storage, Powering the Grid up to 99.9% of the Time," *Journal of Power Sources*, 225, March 2013.

21 SmartGrid.gov, "Overview of Programs, Studies and Activities," www.smartgrid.gov/ recovery_act/project_information (accessed June 10, 2014).

22 Federal Energy Regulatory Commission, Order 792, November 22, 2013; Tina Casey, "Renewable Energy Barriers Fall with New FERC Order," *Clean Technica*, September 27, 2013.

23 "California Passes Huge Grid Energy Storage Mandate," *Greentech Media*, October 17, 2013, www.greentechmedia.com/articles/read/california-passes-huge-grid-energy-stora ge-mandate (accessed June 10, 2014).

24 California ISO, "Energy Imbalance Market," www.caiso.com/informed/pages/sta keholderprocesses/energyimbalancemarket.aspx (accessed September 18, 2015).

25 Wayne Lusvardi, "Energy Imbalancing Market: Bailing Out California Green Power Two Hours/Day," *MasterResource*, November 13, 2013, www.masterresource.org/ 2013/11/energy-imbalancing-market-bailing-out-green-power-two-hours-each-ca-da y/ (accessed June 10, 2014).

26 PricewaterhouseCoopers, *Energy Transformation: The Impact on the Power Sector Business Model*, 13th PwC Annual Global Power & Utilities Survey, October 2013, p. 34.

27 SEIA, *U.S. Solar Market Insight Report*, Q2 2015; EIA, *Annual Energy Outlook 2014*.

28 SEIA, "2014 Top 10 Solar States," www.seia.org/research-resources/2014-top-10-sola r-states (accessed September 15, 2015).

29 *BP Statistical Review of World Energy 2013*, Solar Energy, www.bp.com/en/global/corp orate/about-bp/energy-economics/statistical-review-of-world-energy-2013/review-by-e nergy-type/renewable-energy/solar-energy.html (accessed June 10, 2014); and Solarika, "Top 10 Solar PV Countries – World Overview," October 31, 2013, www.solarika. org/blog/-/blogs/top10-solar-pv-countries-world-overview (accessed June 10, 2014).

30 International Energy Agency, *2014 Snapshot of Global PV Markets*, p. 6.

31 SEIA, *Solar Market Insight Report 2015 Q2*, Executive Summary; "Alternative Energy Will No Longer Be Alternative," *The Economist*, November 21, 2012.

32 See BSR, for example, for a list of corporate members and their Corporate Social Responsibility activities, www.bsr.org/ (accessed June 10, 2014).

33 *High Penetration PV: Experiences in Germany and Technical Solutions*, SMA Solar Technology AG, 2012; "Hawaii's Novel Approach to PV Integration," *Solar Industry*, January 31, 2013, www.solarindustrymag.com/e107_plugins/content/content.php? content.12019 (accessed June 10, 2014).

34 "German Utilities Attack Green Squeeze on Profits, Hint at Leaving," *Reuters*, August 6, 2013, www.reuters.com/article/2013/08/06/germany-utilities-idUSL6N0F V1FX20130806 (accessed June 10, 2014); "German Renewables Drive Hits E.on Profits," Deutsch Welle, www.dw.de/german-renewables-drive-hits-eon-profits/a

-17222326 (accessed June 10, 2014); "How to Lose Half a Trillion Euros," *The Economist*, October 12, 2013.

35 "Utilities Facing a Mortal Threat from Solar," *Wall Street Journal*, March 22, 2013.

36 *Value of the Grid to DG Customers*, IEE Issue Brief, The Edison Foundation, September 2013 (updated October 2013), p. 2.

37 *Value of the Grid to DG Customers*, p. 2.

38 DSIRE, "California Incentives/Policies for Renewables and Efficiency," www.dsir eusa.org/incentives/incentive.cfm?Incentive_Code=CA02R&re=0&ee=0 (accessed June 10, 2014).

39 "Why the US Power Grid's Days are Numbered," *Bloomberg Business Week*, August 22, 2013.

40 *Value of the Grid to DG Customers*, pp. 4–5.

41 Lovins, "Let's Celebrate …"

42 Lovins, "Let's Celebrate …"

43 *Energy Transformation: The Impact on the Power Sector Business Model*, p. 14.

44 *Energy Transformation: The Impact on the Power Sector Business Model*, p. 15.

45 *Value of the Grid to DG Customers*; and *Energy Transformation*.

46 John Finnigan, "The Arizona Public Service Ruling on Solar: Here's Why It's a Win-Win," Environmental Defense Fund, Energy Exchange, http://blogs.edf.org/energyex change/2013/11/26/the-arizona-public-service-ruling-on-solar-heres-why-its-win-win/ (accessed June 12, 2014).

47 Remarks of Ray Mabus, Secretary of the Navy, National Clean Energy Summit 4.0, Las Vegas, NV, August 30, 2011, www.navy.mil/navydata/leadership/mist.asp?q= 505&x=S (accessed June 9, 2014).

48 US Energy Information Administration (EIA), *Annual Energy Outlook 2014*; and EIA, Frequently Asked Questions, www.eia.gov/tools/faqs/faq.cfm?id=27&t=10 (accessed June 9, 2014); and David Bernell, "Ethanol," in Steel, ed., *Science and Politics: An A-to-Z Guide to Issues and Controversies*, Washington, DC: CQ Press, 2014.

49 US Department of Agriculture, National Agricultural Statistics Service, "US Corn Acreage Up in 2011; Soybeans Slightly Down," June 30, 2011, www.nass.usda.gov/ Newsroom/2011/06_30_2011.asp; and Renewable Fuels Association, "Statistics," http://www.ethanolrfa.org/pages/statistics#C (accessed June 9, 2014).

50 David Bernell, "Ethanol," in Steel, ed., *Science and Politics: An A-to-Z Guide to Issues and Controversies*, Washington, DC: CQ Press. 2014.

51 "Using Biofuel Tax Credits to Achieve Energy and Environmental Policy Goals," Congressional Budget Office, July 2010, p. 7.

52 Bernell, "Ethanol."

53 Environmental Working Group, "2013 Farm Subsidy Database," http://farm.ewg. org/progdetail.php?fips=00000&progcode=corn (accessed June 9, 2014).

54 Bernell, "Ethanol."

55 Bernell, "Ethanol."

56 US Environmental Protection Agency, "Renewable Fuel Standard," www.epa.gov/ OTAQ/fuels/renewablefuels/ (accessed June 9, 2014).

57 Renewable Fuels Association, www.ethanolrfa.org/pages/renewable-fuel-standard/, and www.ethanolrfa.org/pages/statistics#C (accessed June 9, 2014).

58 David Murphy, Charles Hall and Bobby Powers, "New perspectives on the energy return on (energy) investment (EROI) of corn ethanol," *Environment, Development and Sustainability*, February 2011; Michael Wang, May Wu and Hong Huo, "Life-cycle energy and greenhouse gas emission impacts of different corn ethanol plant types," *Environmental Research Letters*, 2(2), May 2007.

59 Bernell, "Ethanol."

60 David Pimentel, "Ethanol Fuels: Energy Balance, Economics, and Environmental Impacts are Negative," *Natural Resources Research*, June 2003; Tad Patzek,

"Thermodynamics of the Corn-Ethanol Biofuel Cycle," *Critical Reviews in Plant Sciences*, 23(6) 2004, pp. 519–567.

61 Hosein Shapouri, Paul W. Gallagher, Ward Nefstead, et al., *The Energy Balance of Corn Ethanol: An Update*, US Department of Agriculture, July 2002; Wang et al., "Life-cycle energy and greenhouse gas emission impacts ..."; Michael Graboski, "Fossil Energy Inputs in the Manufacture of Corn Ethanol," Prepared for the National Corn Growers Association, 2002; Marcelo Dias De Oliveira, Burton E. Vaughan, and Edward J. Rykiel Jr., "Ethanol as Fuel: Energy, Carbon Dioxide Balances, and Ecological Footprint," *BioScience*, July 2005; Alexander Farrell, Richard J. Plevin, Brian T. Turner, et al., "Ethanol Can Contribute to Energy and Environmental Goals," *Science*, January 2006; Renewable Fuels Association, *Accelerating Industry Innovation, 2012 Ethanol Industry Outlook*.

62 David Murphy, Charles Hall and Bobby Powers, "New perspectives on the energy return ..."

63 Murphy and Hall, "Year in Review," pp. 102–18.

64 Murphy et al., "New perspectives ..." In the case of oil, the average EROEI over the years has declined. Domestically produced oil in 1930 had an EROEI of 100:1, but this has fallen to closer to 10:1 at present. Still, oil provides a net positive with regard to inputs and outputs. It still "pays for itself." See David Murphy and Charles Hall, "Year in Review—EROI or Energy Return on (Energy) Invested," *Annals of the New York Academy of Sciences*, January 2010.

65 Wang et al., "Life-cycle energy and greenhouse gas ..."

66 Farrell et al., "Ethanol Can Contribute ..."

67 Renewable Fuels Association, *Accelerating Industry Innovation*, p. 25.

68 Farmdoc, University of Illinois Department of Agricultural and Consumer Economics, "US Monthly Average Corn Price," www.farmdoc.illinois.edu/manage/uspricehistory/us_price_history.html (accessed May 14, 2015).

69 Timothy Wise, "The Cost to Developing Countries of U.S. Corn Ethanol Expansion," Global Development and Environment Institute, Working Paper No. 12–02, October 2012; Donald Mitchell, "A Note on Rising Food Prices," *The World Bank, Policy Research Working Paper* 4682, July 2008.

70 Index Mundi, "Commodity Prices," www.indexmundi.com/commodities/ (accessed June 9, 2014).

71 Marco Lagi, Karla Z. Bertrand, and Yaneer Bar-Yam, "The Food Crises and Political Instability in North Africa and the Middle East," Cornell University Library, *arXive.org*, August 11, 2011, http://arxiv.org/abs/1108.2455 (accessed June 9, 2014).

72 Lagi et al., 2011.

73 Bernell, "Ethanol."

74 *USA Today*, "Ethanol Pumping Up Food Prices," February 9, 2011, http://usatoday30.usatoday.com/money/industries/food/2011-02-09-corn-low_N.htm (accessed June 9, 2014).

75 Deepak Rajagopal, S. E. Sexton, D. Roland-Holst, and D. Zilberman, "Challenge of biofuel: filling the tank without emptying the stomach?" *Environmental Research Letters*, November 2007.

76 *2013 Ethanol Industry Outlook*, Renewable Fuels Association, p. 10.

77 "Using Biofuel Tax Credits to Achieve Energy and Environmental Policy Goals," Congressional Budget Office, July 2010, p. 2.

78 Al Jazeera, "Corn Lobby Outgrows US Farm Subsidies," August 31, 2012, www.aljazeera.com/indepth/features/2012/08/2012826114433916589.html (accessed June 9, 2014).

79 Iowa Corn, "Iowa's Corn Production," www.iowacorn.org/en/corn_use_education/faq/ (accessed June 9, 2014).

80 "Clinton Meets with Ethanol Representatives," *The Hill*, April 15, 2015, http://thehill.com/policy/energy-environment/238953-clinton-meets-with-ethanol-representati

ves (accessed May 14, 2015); "U.S. Republican Hopefuls Bush, Walker Change Their Tune on Ethanol," *Reuters*, March 7, 2015, www.reuters.com/article/2015/03/07/us-usa-politics-iowa-idUSKBN0M30UT20150307 (accessed May 14, 2015).

81 Konrad Imielinski, "Ethanol—Tracking the Presidential Candidates: John McCain," *Seeking Alpha*, March 14, 2007, http://seekingalpha.com/article/29592-ethanol-tracking-the-presidential-candidates-john-mccain (accessed June 9, 2014).

82 "Iowa Caucuses Show Ethanol No Longer Sacred to Rural Voters," *Bloomberg.com*, January 4, 2012, www.bloomberg.com/news/2012-01-05/iowa-caucuses-show-ethanol-no-longer-sacred-to-rural-voters.html (accessed June 9, 2014).

83 Environmental Protection Agency, "Notice of Decision Regarding Requests for a Waiver of the Renewable Fuel Standard," *Federal Register*, 77(228), November 27, 2012, p. 70754, www.epa.gov/otaq/fuels/renewablefuels/notices.htm (accessed June 9, 2014).

84 Environmental Protection Agency, "EPA Decision to Deny Requests for Waiver of the Renewable Fuel Standard," November 2012, www.epa.gov/otaq/fuels/renewablefuels/notices.htm (accessed June 9, 2014).

85 EPA, "EPA Proposes 2014 Renewable Fuel Standards," Regulatory Announcement, November 2013.

86 Renewable Fuels Association, "Blend Wall: Big Oil Builds the Blend Wall," www.ethanolrfa.org/pages/big-oil-builds-the-blend-wall (accessed September 17, 2015).

87 M.R. Schmer, K.P. Vogel, R.B. Mitchell, and R.K. Perrin, "Net energy of cellulosic ethanol from switchgrass," *Proceedings of the National Academy of Sciences*, January 15, 2008.

88 EPA, "EPA Issues Direct Final Rule for 2013 Cellulosic Standard," Regulatory Announcement, April 2014.

6

THE INTERNATIONAL DIMENSIONS OF US ENERGY SECURITY

The world is fraught with dangers. While the specific risks have changed, history illustrates well that risks have and will likely always abound. There are numerous conflicts going on around the world that pose risks for energy supplies and prices. In the Middle East, Iran and the United States have signed an agreement on Iran's nuclear program; ISIS controls parts of Iraq and Syria and is making advances; a coalition in the Middle East being led by Saudi Arabia launched airstrikes in Yemen in 2014 and is seeking to oust the Iranian-backed Houthis, who control the country's capital. In Europe Russia continues to be at the center of crisis in Eastern Europe through the support of separatist groups in Ukraine, while holding the threat of disrupting natural gas supplies. And in Asia, China has begun constructing an airfield and other infrastructure in the disputed Spratly and Paracel Island chains in order to bolster its claims of ownership to the uninhabited territories.

The attempt in this chapter is not to thoroughly examine each of these risks and crises with respect to energy, to solve global energy problems, or to employ a historical analysis to lay blame, explain current crises and provide definitive lessons learned. If we did learn from history, then energy consumers around the world would be likely to dramatically and permanently reduce dependence on far-flung energy and natural resource markets for fuels that cause lasting environmental harm. The key energy resource—oil—that the world pursues vigorously is finite in supply (though new discoveries and technology advances continue to allow global supplies to meet global demand) and faces growing demands as developing countries become richer. This, in itself, creates enormous energy security risks to those who maintain long-term reliance on petroleum. International sources of supply will eventually disappear or become nearly unmarketable due to the cost of resource extraction and its transportation to distant markets.

Consider this: if the United States could start with a clean slate today and have an opportunity to choose from a variety of energy sources, it seems highly unlikely that there would be great support for massive reliance upon an energy source that has to be shipped to the domestic market from around the world, that is subject to periodic disruptions resulting from political conflict in the Middle East, that is characterized by great volatility in prices, that requires large investments of money, energy and water to produce for consumption, that pollutes the air and changes the global climate when used as intended, that periodically spills and causes large and small bodies of water to become polluted, and that will eventually run out, or more accurately, will become too expensive to access once the most accessible reserves have become depleted.

Rather than considering these "what-ifs," this book examines the energy wants and needs of the United States and highlights the dilemmas and risks inherent in attempting to meet a plethora of multifaceted energy security goals. To that end, the focus in this chapter is on key global issues *within the context of US policy concerns and actions*. Emphasizing again the central argument of this book—that the effort to achieve greater energy security contains its own contradictions and obstacles—we consider a framework for understanding US energy security objectives and energy policy with regard to the rest of the world, and how a number of global institutions and actors impact the energy security of the United States. While US energy policy in the domestic realm has come to be about moving toward alternative, cleaner renewable energy resources, expanding domestic supplies of oil and natural gas, reducing energy demands, and increasing efficiency, the global dimensions of US energy policy are generally about trying to find a better way to ensure stable, reliable, global markets and supply chains for oil and natural gas, while diminishing the potential for energy-related political and military conflict, or at least, insulating the country as much as possible from the adverse impacts of such conflict. Of course, these goals are also tied in part to the question of how much oil and gas can be produced domestically.

A Framework for the US View of Energy Security

On August 1, 2007, sailors from two Russian mini-submarines planted a titanium Russian flag on the seabed of the Arctic Ocean, 14,000 feet under the ice at the North Pole. As a technological feat, it was a potentially dangerous mission and a record-breaking dive, even if in practical terms it meant little for anything the Russians are currently doing in the Arctic. As a political matter, it was significant, and potentially troublesome, as the Russians were making a political claim, whereby planting their flag was connected to the assertion of a right to exploit natural resources in the area.

Establishing a sovereign right to natural resources in a maritime setting is nothing new or surprising. Countries do this all the time in their territorial waters, and in areas where they are granted economic rights under international law. In waters not globally recognized as part of a particular country's territory, this type of action can be troubling, and it is increasing. Countries such as the United States and Canada were quick to dismiss the Russian move as a stunt. As the Canadian Foreign Minister said, "This isn't the 15th century. You can't go around the world and just plant flags and say 'We're claiming this territory'."[1] But Russia has made the claim that a portion of the Arctic seabed is Russian maritime territory, and that the vast energy resources there are thus Russia's.

If this type of political conflict over energy resources is a sign of things to come, it represents a worrisome development. One of the great benefits the United States and its allies achieved in the wake of World War II was the restoration and strengthening of an open, global trading regime—a set of rules that made it possible for countries to acquire the resources they wanted by trading, rather than being tempted to seize or conquer lands with resources they coveted. One of the lessons of the war, in which Japan and Germany pursued (among other things) economic strength and self-sufficiency through war and conquest, was most certainly that there is great value in building and maintaining a liberal trading order.

This type of approach to the acquisition of goods has been a consistent feature of US foreign policy, which has long seen US security as being tied to a maritime-based, global trading system. In the 1940s, with the adoption of the General Agreement on Tariffs and Trade (GATT), the US took its preferred way of doing business global, leading a global effort to move countries away from the imperial policies and practices that led to European colonial control over most of the world. In the era of the GATT, and its successor, the World Trade Organization (WTO), this liberal economic order has been remarkably successful in helping to prevent violent conflict over natural resources, at least among the world's big powers. During this period, trading rules and institutions, along with US power and leadership, helped to break, or at least greatly diminish, the links between nationalism, violence and resource sufficiency. In the postwar period, in order for the US, or Japan, or Europe to get "their" oil, they have not needed or wanted to conquer or colonize lands that held the oil. Getting to this point has been a key goal of US global energy policy, encompassing an overarching framework for the US energy security worldview, namely the de-politicization and the de-militarization of energy markets, which particularly impacts oil and natural gas.

The US certainly has a long record of dealings in order to install or support friendly governments, whose energy policies have conformed to American goals and interests. The ousting of the government of Iran in 1953 is a notable example of this. More recently, the US-led war in 1991 to remove Saddam Hussein's Iraq from Kuwait, and push back Iraq lest it become too dominant in the region and control such a large share of the world's petroleum reserves, was indeed a massive and violent undertaking. These actions certainly helped to sustain US power, as well as control over how the global oil market would operate.[2] Nonetheless, these examples are a far cry from Japan's Greater East-Asia Co-Prosperity Sphere of the 1930s and 1940s, in which Japan sought to conquer and control most of

East Asia and the Western Pacific to ensure its economic security and self-sufficiency. The US has not sought outright control of Iranian or Iraqi land and oil resources—for itself or for US oil companies—in a way that Japan did in East Asia. The levels of violence, insecurity, destruction and disorder between the two eras are incomparable. One may argue that the pre-World War II period was a different world, but that is exactly the point. It was. Now, the US and others can largely rely on markets to meet their energy needs, and even if the US periodically favors one government over another in a particular country, such favoritism is, according to a US-centered viewpoint, aimed at ensuring countries operate within the trading rules and international norms that govern a more liberal trading and economic order. US policy by this formulation has not been designed to ensure exclusive American access to oil resources, or even to assure that contracts go to American companies. Instead, its broad objectives have largely been aimed toward assuring that a global energy market can continue to function. In other words, it uses policy and state power to allow a market to allocate oil resources, rather than using policy and power to supplant a global oil market. According to this understanding, the United States characterizes its policy as one aimed not only at promoting its own interests, but a larger, general interest that benefits others (though it is often the conceit of the powerful to assume that other countries do in fact, or at least should, share their interests).

With respect to energy, this system may be starting to show signs of breaking down. The demand for oil and natural gas around the world is growing, and is expected to do so for a long time. The US Energy Information Administration projects that global oil demand will rise from 87 million bbl/day to 119 million bbl/day by 2040. With respect to natural gas, demand is expected to rise from 118 to 185 trillion cubic feet in the same timeframe.[3] In the efforts states are making to ensure that they will have sufficient and reliable energy supplies—and this means largely oil and natural gas—to fuel their economies, it appears that nationalism is becoming increasingly tied to the problem of resource sufficiency. Countries, including the United States, want "their" oil and natural gas, and are increasingly seeing energy acquisition as a zero-sum game. This is what drove the Russian expedition to the North Pole, a development that Michael Klare calls "the race for what's left," and Britain's *Daily Telegraph* characterized as the beginning of "the world's last colonial scramble."[4] It is what drives America's goal of seeking to deny regional power and influence in the Persian Gulf to other major powers, while trying to bolster its own. It is what drives China's vast undertakings in Africa and Central Asia to secure energy resources through state-to-state contracts and direct pipelines. Unfortunately, this trend seems likely to increase.

Oil-Fueled Conflict

If the United States, China and others seek to acquire and use the massive quantities of the same fossil fuels the world has used for the last 100 years, and to consume energy at the level they expect and want, while at the same time adding to the mix the nationalistic fears and assertions that are likely to increase as a result of energy insecurity, it means that the diverse and sometimes competing energy security objectives we have identified—meeting energy needs, procuring energy economically, protecting national security, and protecting the environment—will clash with one another. To simply seek out more oil and natural gas to meet growing needs means that the threats both domestically and abroad will be likely to multiply.

US foreign policymakers identify a number of potential threats to global and US national security resulting from the worldwide dependence and increased use of oil. First of all, there is the problem of nuclear proliferation. The number one example of this from the US perspective has been Iran, which benefits greatly from high oil prices and has been able to allocate resources toward its nuclear program. Though the government of Iran has protested that its nuclear program is not meant for the development of weapons, such claims have been unconvincing to many other countries, particularly the United States. This has led to framework agreement arrived at in 2015 by American and Iranian negotiators; it has led the US and Israel to consider the possibility of a military strike to prevent Iran from developing a nuclear weapon in the absence of agreement or failure of implementation; and it leads to concerns about a regional arms race, as neighboring states may consider building their own nuclear weapons in response to any Iranian gains. What's more, Iran has enjoyed a level of protection from attack due to its ability to adversely impact the oil market by threatening the Strait of Hormuz. This "oil shield" means that if the US, Israel or others seek to protect their security interests regarding nuclear proliferation through military action against Iran, they could very well harm their own, and the world's, economic interests at the same time.[5]

Another consequence of the heavy reliance on oil usage can be seen in the level of increased violence and terrorism from non-state actors in the Middle East. This has come in many forms. Iran has provided longstanding support to several organizations in the Middle East, such as Hamas and Hezbollah, who have carried out violent attacks against Israel. It is expected that such organizations would be even more emboldened by an Iran with nuclear weapons capabilities. Terrorism also comes indirectly from other oil-rich states such as Saudi Arabia, where some portion of the petrodollars supplied by the US and other Western consumers finds its way into the coffers of organizations such as al-Qaeda that are hostile to the US and its allies. Petrodollars, at least indirectly, helped to pay for the attacks of 9/11, as well as ongoing violence by non-state actors in the Middle East.[6] In addition, the fallout from US intervention in Iraq has led to the rise of ISIS and its brand of brutality, violence and terrorism in Iraq and Syria. The organization's ability to control territory and people while maintaining its fighting capacity is due to the fact that it finances its operations predominantly from the sale of oil (it is also funded by private donors, "taxation" and extortion of captive populations, seizure of assets in territories it controls, the sale of plundered antiquities, kidnapping ransoms, and the sale of women and girls in the sex trade).[7]

This is only part of the list of threats the United States can identify that tie to America's and/or the world's energy usage. There are many other circumstances that the US would rather avoid, though in some cases, the US itself is a contributor to the problem it is seeking to address:

- Russian influence over former Soviet Republics and the rest of Europe. The taking of Crimea and the conflict in eastern Ukraine have taken place amidst the renationalization of the energy sector in Russia, and attempts by Russia's leadership to reestablish its global influence through control over energy resources. To that end, Russia has used energy as a tool of its foreign policy. The energy giant Gazprom, of which 51% is owned by the Russian government, has a relationship with the state such that "Gazprom has become the Kremlin's faithful servant and dangerous sword, and the Kremlin has become Gazprom's protector."[8] Russia, via Gazprom (one of the world's largest energy producers), has repeatedly cut off gas supplies to Ukraine and other parts of Europe, it has sought to gain control over the gas transportation infrastructure that used to be part of the Soviet Union's gas supply system, and it provides price discounts to countries loyal to Moscow.
- The growing assertiveness and assault on personal liberties in states like Russia, Iran and Venezuela, which earn most of their income from oil. In these states, rulers have ensconced themselves by building up their security forces and taking greater control of the media, while harassing, jailing and silencing their critics. At the same time they use their oil revenues to buy off and divide any potential opposition by providing subsidies and benefits to citizens and interest groups. As Thomas Friedman of *The New York Times* has described this relationship, a trip to the gas station is effectively saying, "fill 'er up with dictators."[9]
- Competitive arms transfers and military assistance. Oil consumers have for years provided military assistance to oil producers. The US has done so with Saudi Arabia for decades, and has in recent years done so among several Central Asian republics. China is the newest entrant into this type of relationship, providing arms to several countries in Africa and Central Asia. It has even been accepting oil and minerals in lieu of cash to pay for military equipment.[10] These relationships strengthen ongoing relations, win favor with suppliers, and help to secure favorable terms in supplying oil.
- Venezuela's pursuit under President Hugo Chavez, and his successor Nicolas Madura, of an anti-US bloc in Latin America and the Caribbean. The Bolivarian Alliance for the Peoples of Our America, or ALBA, organized and funded by Venezuela, is an association of eight countries seeking a level of regional economic integration and ideological solidarity that stands in direct opposition to US power and influence in the Western Hemisphere.

- Looking the other way at human rights violations. Oil exporters like Azerbaijan, Kazakhstan, Angola and Sudan receive support at the UN and/or avoid probes into their violations of human rights.

The larger context to all these developments is that global dependence on oil increases resource competition as demands grow. The United States and China have already experienced one instance of this when the Chinese company CNOCC sought to purchase the American oil company Unocal in 2006. The outcry in the US Congress was so swift and loud that the sale was abandoned, but the example serves to underscore the nationalistic feelings associated with securing energy supplies. In places throughout the world, resource competition occurs with growing frequency. There are numerous claims to oil resources in parts of the Arctic that Canada asserts are part of their territorial waters, and by other countries who claim the undersea resources that lie in international waters. China and its neighbors have made competing claims to the Spratly Islands in the South China Sea. And Japan and China also have competing claims to the Chunxiao gas fields in the East China Sea, which China has begun to develop. This dispute was exacerbated by Japan's seizure of a Chinese fishing boat in 2010 that was operating in waters controlled by Japan but also claimed by China as well. The fishing boat incident prompted China to cut off shipments of rare earth metals to Japan for a period of time.

BOX 6.1 AN ALTERNATIVE TO THE US VIEW

The American view of global energy security, and its understanding of what constitutes politicization and militarization of global energy markets, is not necessarily shared around the world. In an era when the United States enjoyed greater power and influence around the globe than it does today, divergence with the American framework may not have been of great consequence. However, this challenge to the US is of particular importance in a context of rapid global change, particularly as China, the world's fastest rising power, takes issue with the United States and its energy policies.

An article in the *OPEC Review* in 2007 by Wu Lei and Liu Xuejun entitled "China or the United States: Which Threatens Energy Security?" encapsulates this critique, arguing that the US is the more significant threat to global energy security. The authors argue that the US has a preeminent priority in the promotion of energy security for itself and its allies, but that the US often acts in ways that work against its own stated priority and, in the process increases global energy insecurity in terms of oil supplies. Reductions in supply along with underinvestment in resource development are largely due to US-led sanctions and embargoes over several decades, having targeted major producers such as Iran, Libya, Iraq, and Sudan. "The biggest energy security problem America faces is the contradiction between its foreign policy and its energy-related foreign policies ... the country has every reason to put energy security at the top of the nation's political agenda. However, careful observations lead us to believe that other concerns, rather than energy security, are prioritized."[11]

If the US is concerned over projections that China will consume half of the world's petroleum by the mid-21st century, it has conveniently ignored that US policy has greatly contributed to this worry.

The authors also rebut the often-expressed US narrative that Chinese oil companies are "locking up" world oil supplies. Using data from the US Energy Information Administration, the authors point out that Chinese investment in Middle East and African oil and natural gas fields is eclipsed by US energy firms' investments. US energy giants are much larger than Chinese national oil companies. Additionally, they note that China still purchases a large percentage of its fossil energy on the open market, not as "equity oil" that flows directly to China from a producer nation as a result of Chinese investments in oil field development.

Lei and Xuejun also point out that Chinese national oil company ventures outside of China are not solely an extension of national government strategy, nor are they designed to threaten US interests. As China has moved to a more corporate for-profit model, Chinese oil companies must act as a for-profit business, seeking out new resources as energy demand grows, as well as identifying and investing in lucrative energy business opportunities throughout the world. At times, this means investments in nations with regimes that the US does not support. However, Chinese government and business interests should not be measured by whether or not they follow US interests. China must pursue its own interests as it sees them.

Doug Stokes and Sam Raphael, in *Global Energy Security and American Hegemony*, extend this critique of US energy security policy. Far from seeking to prevent the politicization and militarization of energy markets, they argue, it is the United States itself that has introduced and sustained such militarization. Successive administrations have deployed military force, in such places as Iraq, Kuwait and Saudi Arabia, and "most often in the guise of counterinsurgency training and equipping to friendly security forces—in order to stabilize oil-rich zones ... with the aim of disciplining those social forces that have the potential to organize against US oil hegemony." To that end, "[by] acting as the ultimate guarantor of global energy security, US hegemony over the international system is consolidated, with potential rivals to its position forced to be (and in some cases content to be) reliant upon American power."[12] Far from being a disinterested arbiter of open energy trading, the United States is, in this formulation, cementing its power at the expense of others.

Domestic Production of Fossil Fuels and National Security

Many of the challenges the US faces abroad have a link, directly or indirectly, to the worldwide reliance on oil and natural gas. To the extent that the United States is able to add to the global supply through increased production of both commodities at home, can this diminish the national security concerns noted above? What these newfound energy sources may mean in the long term for US troop deployments overseas, the assertiveness of oil-rich dictatorships, the threat of terrorism, and other national and global security concerns remains to be seen, but there is little question that the ups and downs of the global oil market, along with the actions taken by nations around the world in response to this market, have been greatly influenced by the American energy boom. It is difficult, if not impossible, to prove a negative, but it is not unreasonable to think that political instability and active hostilities in Iraq and Syria, upheaval and state failure in Libya, and concerns over Russia's use of energy as a political "weapon" would be putting upward pressure on global oil prices in the absence of increased US production of oil and natural gas. After all, in the midst of all these conflicts, the increase in US petroleum production, after several years of growth, had the effect in 2014 of pushing oil prices downward very quickly. They dropped by more than half from June 2014 to January 2015, from over $100 to $47 a barrel in this timeframe.[13]

The growth of unconventional oil and gas production in the US has led to a changing understanding about how energy security can potentially be enhanced. Diminished reliance on foreign suppliers of oil, along with the potential to export natural gas, suggest that the United States could be better insulated from disruptions in energy supply chains that stretch around the world, while also reducing the global impacts of such disruptions. In other words, this is expected to have the effect of enhancing the abundance, reliability and affordability of energy supplies, while reducing national security vulnerabilities. Moreover, this type of "energy independence" (at best a misnomer, or at least a term that means different things to different people) has been a longstanding element of US energy policy. When one considers US military involvement in the Middle East ever since the pronouncement of the Carter Doctrine in 1979, and the adverse impact of conflict in the region on energy supplies and fuel prices, the benefit is not insignificant.

Since the oil embargo of 1973, the American view has held that the national security vulnerabilities inherent in US dependence on oil are vast. This view was clearly articulated in a report issued in 2006 by the Council on Foreign Relations entitled *National Security Consequences of U.S. Oil Dependency*.[14] The task force that wrote the report was chaired by John Deutch, who had served as CIA Director, Deputy Secretary of Defense, and Undersecretary of Energy; and James Schlesinger, who had served as CIA Director, Secretary of Defense, and Secretary of Energy. If any publication could reflect the American energy security worldview, this was it. Released in 2006, when US forces were mired in Iraq, the country was importing two-thirds of its daily petroleum needs, and oil prices were around $80/bbl and on the rise, the report pointed out the following problems (which had a heavy focus on Iran):

- "The Persian Gulf ... is a region that has been unstable and includes countries that have periodically used their oil exports for political purposes unfriendly to the United States."
- "Dependencies in the world oil market empower oil rich countries—such as Iran and Venezuela—to carry out foreign policies that are hostile to that of the United States."
- "Concern about losing Iran's 2.5 million barrels per day of world oil exports will cause importing states to be reluctant to take action against Iran's nuclear program."
- "France and Germany, and with them much of the European Union, are more reluctant to confront difficult issues with Russia and Iran because of their dependence on imported oil and gas."
- "A large fraction of the world's traded oil already passes through a handful of strategic choke points, such as the Strait of Hormuz."
- "The infrastructure for delivering oil has several potential weak links, including major oil processing facilities that are vital yet vulnerable to attack and difficult to repair."
- "Payments for oil lead to large dollar balances built up by oil producers, 'petrodollars,' giving them potential leverage over U.S. capital markets."[15]

With respect to every one of these points, the growth of US oil production offers the opportunity to mitigate the vulnerabilities, which can be seen as far less threatening from a US perspective than they did in 2006. With respect to natural gas, while the anticipated benefits to the United States may not as pronounced as those with oil, they have still been seen as significant. Examples of this optimism abound. President Obama has touted his administration's efforts to "strengthen our own energy security" so that "today, the number-one oil and gas producer in the world is no longer Russia or Saudi Arabia. It's America."[16] Former State Department official Carlos Pascual, commenting on US efforts to persuade other countries to place sanctions on Iran, explained in 2014 that "we were able to have a different kind of conversation than we could have had even 10 years ago," and noted the value of pointing out "what the trends were in [oil] markets, where supply was coming from—in particular, from the United States."[17] In the media, *The New York Times* reported that "Oil's Comeback Gives U.S. Global Leverage," while *US News and World Report* captured the sentiment in the story "An Oil Boom Is a Power Boon."[18]

An increased level of oil production and the sense of optimism it has aroused does not mean that the United States is immune to the actions of other countries.

There is a global oil market, not a series of insulated regional markets. Therefore supplies, prices and market activity will ultimately reflect the cumulative actions of all market participants. Moreover, the expected and unanticipated political and economic consequences of sustained increases in US fossil energy supplies could be numerous, and not always advantageous to US interests: economic contraction and political instability in oil-producing states, including countries both friendly and unfriendly to the United States; greater Chinese influence in the Persian Gulf as it comes to purchase a larger share of the oil produced in the region; closer ties between China and Russia as sales of Russian oil and gas to China produce a greater confluence of interests; decreased production of oil overseas, producing little net reduction in market sensitivity to disruptions in oil supplies.

Notwithstanding these concerns, the changes in the fundamental geopolitical energy supply dynamics for the United States have prompted a whole new debate over energy security policy. If the United States can increasingly be conceived of as an energy producer nation, and not only a massive consumer of energy, to what extent should the United States be making these resources available for export on the global market, and what would be the impacts on national and global security of greatly enhanced exportation? In other words, does energy security derive more from open, global markets, in keeping with the larger thrust of US trade and foreign policy described above, or from keeping energy resources at home? US law prohibited the export of crude oil for decades, though the wisdom of this policy (enacted in 1975 as part of the *Energy Policy Conservation Act*) came into question in the midst of the shale oil and gas boom. In 2015 Congress approved legislation to lift the ban and President Obama signed the measure into law.

The prohibition on energy exports did not apply to natural gas, and the industry is moving quickly to capitalize on this opening for export production and trading. Considering that natural gas traded in mid-2015 in the United States at near $3 per MMBtu, $7 in Europe, and $9 in Japan, the enthusiasm to export is understandable.[19] To that end, several projects are underway to build terminals for the export of liquefied natural gas (LNG). As of September 2015, the Federal Energy Regulatory Commission had approved plans to build five LNG export terminals in Texas, Maryland and Louisiana (though not without noteworthy local opposition), and had proposals for an additional twenty-two terminals.[20] In spite of the industry's move toward natural gas exports, there is a strong argument to be made for keeping domestic supplies off the global market, reserving them for future use at home. This was the logic applied to the crude oil export ban. According to this formulation, as the country is expected to have a longstanding need for natural gas, it makes good sense to reserve a steady supply for domestic use well into the future. The United States remains a net importer of natural gas, and it is unclear if the high expectations of increased production will continue to be met in the long term. In an era of what looks like increasing global competition to secure access to energy sources, the shale revolution means that the US has more of its own oil and gas, secure within its borders, providing greater reliability. Exporting natural gas, by this logic, is perhaps acting in a short-sighted way. Besides, when one considers the fact that making US shale available on the global market will raise the domestic price of natural gas, it appears that one outcome of substantial exportation will be to enrich energy companies at the expense of American consumers.

However, there is another set of considerations to be taken into account with respect to national and global security. In an era of "the rise of the rest" and the rapid economic development of historically poorer countries, the demand for energy is growing, and investors and countries that have attracted support for growth are seeking access to new oil and gas supplies. The result has been more state-led activity in seeking access to energy, or pursuing a "resource nationalism" course of action, clearly a potentially volatile circumstance. For example, in recent years Russia has made claims upon Arctic resources, jockeying for access. Japan and China have made rival claims upon uninhabited islands in the South China Sea in order to access energy resources there. And China has concluded oil and gas contracts with both Russia and Kazakhstan, bypassing typical market mechanisms to buy and sell energy in state-to-state contracts, to give but a few examples. The conflict between Russia and the Ukraine, in which natural gas supplies and prices since 2006 have been used as a tool of Russia's foreign policy, further demonstrates the risks associated with the politicization of energy markets. The caution exhibited by the EU countries in dealing with the annexation of Crimea by Russia and the shooting down of a Malaysian airliner with European passengers over Eastern Ukraine in 2014 illustrate the great geopolitical value of Russian gas reserves vis-à-vis the rest of Europe.

When one considers the general American approach to trade and markets throughout its history, a decision not to participate in certain elements of global energy markets by withholding supplies for domestic use seems incongruous. As discussed earlier, the United States has long seen its security as being tied to an open, global-scale trading system. This liberal economic order has been understood, as a matter of US policy, to represent a success in preventing great power conflict. In the context of the debate over whether or not to encourage oil and natural gas exports, applying this same logic means that the United States should want to encourage exports, and ultimately promote a more unified global market within which countries in need of supply can access countries with excess capacity. The more the market for energy involves an open, global system, the more difficult it is for individual countries or OPEC to be successful in controlling supplies and prices, much less initiating and sustaining an energy embargo.[21]

Currently, the world is largely divided into three major regional natural gas markets. These are in North America, Europe and Asia. By contrast, a world consisting of multiple suppliers, a plethora of LNG terminals, and a system in which sellers can ship natural gas virtually anywhere in the world, and buyers can choose to purchase from among a large number of suppliers, seems much better suited to diminishing the potential for political conflict over energy resources. As John Deutch argues in an essay called "The Good News about Gas," a "transparent global gas market would be beneficial to Washington and its allies because it would lead to the economically efficient use of resources and prevent major exporters from exploiting their resources for political gain." To that end, says Deutch, "the United States should refrain from erecting barriers to the natural gas trade, by imposing either import or export duties. If it adopts protectionist measures, other countries will be encouraged to do so too."[22]

This type of argument enjoys strong support among the US foreign policy establishment. A task force co-chaired by Ambassador Paula Dobriansky, Governor Bill Richardson, and Senator John Warner issued a report that examined the national security implication of US shale oil and gas production. The report states that "as the United States imports less energy, some policy leaders hope that a push toward energy isolationism will insulate the country from instability." However, the report concluded, "Such hopes are unfounded. Hoarding energy at home, neglecting bilateral relationships with major global energy players and forfeiting economic opportunities to export energy would leave the United States less secure."[23]

This argument took on even greater salience in 2014. Russia's takeover of Crimea in 2014, its increase in natural gas prices for the Ukraine, its statements about protecting Russian-speaking minorities in former Soviet republics, and the ongoing uncertainty about its intentions and actions, along with Europe's reliance on Russian natural gas, all prompted a discussion about the potential value of US exports to make Europe less dependent upon Russian supplies, to effectively take away one of Russia's major regional political advantages.

Global Institutions

The domestic production of oil and natural gas is one side of the equation in thinking about the global dimensions of US energy security. At the same time, US policy has also sought to shape the international context with respect to the operation of energy markets. Considering the US global energy security worldview, and the recognition that countries will necessarily compete in global energy markets—as either consumers, producers, or both—the United States has been supportive of several efforts to promote international energy cooperation and coordination. At the same time that OPEC considers production levels among members, Russia threatens natural gas supplies to Ukraine, ISIS threatens to take control of oil fields in Iraq, the United States maintains a large military presence in the Middle East, and China enters into agreements worldwide to import oil and natural gas, there are simultaneously a number of important international governing bodies that endeavor to deal with global energy security in a multilateral framework, sharing a common objective of promoting global governance with respect to energy.

The Value of Global Institutions

There can be great value to global governance, which involves the establishment of institutions, rules, procedures and norms to create venues for channeling disputes and encourage cooperation and agreements. Even in the absence of formal enforcement mechanisms, multilateral institutions can still positively affect state behaviors. This is one of the benefits of international organizations and law in a broad sense. Such institutions may not dictate a nation's actions, but they can be helpful in changing or moderating behavior. This point can be missed when considering international law as similar to criminal law. In criminal law, the key question is "how is it enforced?" However, international law, customs and norms are better understood as more analogous to contract law, which is designed to facilitate agreement. In contract law, a key question, which is particularly relevant in the global context, is "does one pay a price for violating it?" And countries can be made to pay a price—sometimes only a small one—for violating the rules and norms of global institutions. This often comes in the form of withholding cooperation in other areas in which a country values assistance, support in a dispute, or collaboration. The WTO does an excellent job of this, allowing for retaliatory tariffs by countries suffering from another country's unfair trade practices. Thus, countries may have more to gain from abiding by the terms of international organizations than by trying to maneuver around them.

In a broader sense, as seen from the liberal internationalist theory of global politics, multilateral agreements and institutions can be seen as central to the establishment, maintenance, and success of the contemporary international system. The "dense network" of institutions, treaties, trade agreements, rules, procedures, laws, norms and customs is an inherent part of the very fabric of the global system. As John Ikenberry argues:

> The liberal international order is not just a collection of liberal democratic states but an international mutual-aid society—a sort of global political club that provides members with tools for economic and political advancement. Participants in the order gain trading opportunities, dispute-resolution mechanisms, frameworks for collective action, regulatory agreements, allied security guarantees, and resources in times of crisis.[24]

With regard to energy security, this view argues for strengthening the global governance regime.[25] Andreas Goldthau and Jan Martin White, in *Global Energy Governance: The New Rules of the Game*, point out that even though public policy debates are characterized by "lopsided attention to the geopolitical dimension of energy security based on the myopic and erroneous presumption that global energy politics is necessarily a zero-sum game," there is also an important "international architecture that underpins global energy ... via financial markets, investment treaties and trade agreements."[26] Making full use of this network of energy institutions and rules is therefore of great value. As Dimo Bohme argues in *EU–Russian Energy Relations: What Chance for Solutions?*, "the essentially global problems in the energy domain need global solutions, which go beyond the prevailing approaches of risk reduction, crisis management and geopolitics," and a "multilateral institutional framework appears to guarantee best a cooperative spirit necessary to reduce physical, economic, environmental and social risks related to energy consumption and external dependencies on energy."[27] Thijs Van de Graaf and Kirsten Westphal suggest adding to this institutional capacity. In spite of a plethora of global energy organizations, they argue for the value of the G8 and G20 in the energy realm, as they can be enlisted to serve as "Global Steering Committees" for energy governance. They argue that:

> The advantage of these "minilateral" clubs is that they bring together the leaders of a small number of key countries in an informal setting. As long as large countries are reluctant to transfer substantial authority over energy issues to formal multilateral settings, such high-level forums seem well placed to steer the global governance efforts with regard to this highly strategic and complex policy issue. Having no fixed agenda, the G-clubs are perhaps the only global forums where countries can discuss the grand objectives of global energy governance in an integrated way.[28]

These understandings of global energy governance have been put into practice repeatedly, as there are several prominent examples of international energy institutions in operation. Among these that we examine are the International Energy Agency (IEA), the Organization for Economic Cooperation and Development (OECD), the Energy Charter Treaty (ECT), and the International Energy Forum (IEF). In all four cases, these organizations do not have the same membership (though there is plenty of overlap), and they have different missions and varying goals. We consider them here to assess their role in supporting the attainment of energy security.

International Energy Agency

The International Energy Agency (IEA) is the global institution that represents the energy concerns of the world's richest, most economically developed nations. It was founded in 1974 in response to the oil embargo of 1973–4, which produced an energy supply shortfall and price spike. More broadly, it represented a response to the power of OPEC, which had been clearly demonstrated in the embargo, and its ability to advance the interests of petroleum exporting nations at the expense of energy importing countries. Established within the framework of the Organization for Economic Cooperation and Development (OECD) as an autonomous agency focused entirely on energy, the agency's original goal was to work with member countries to develop collective responses to oil supply disruptions. This meant reducing reliance on foreign energy supplies through reduced energy demand and conservation, along with the stockpiling of reserves to be used in emergencies. Such emergencies have prompted the release of oil by the US three times: in 1991 leading up to the Gulf War; in 2005 after Hurricane Katrina, and in 2011 when war in Libya disrupted supplies. The IEA has expanded its mission since its founding to what it calls its "four main areas of focus: energy security, economic development, environmental awareness, and engagement worldwide."[29] Only OECD members can join the IEA, but not all OECD members are part of the IEA. States must meet certain qualifications to be included, such as having a 90 day reserve of oil for net importer nations, an oil demand reduction program to limit consumption by up to 10%, legal authority and administrative capacity to enact specified emergency response measures, and reporting requirements to ensure transparency.[30]

The IEA serves as a key collector of information, gathering vast amounts of data from member states on energy availability, development, production, use, imports and exports. It produces reports and analyses on energy trends and projections, and it advocates for strengthening environmental sustainability goals among its member states. The agency also serves as an important coordinating body, working with member and non-member states to provide information and coordinate efforts. Given its origins, it is no surprise to find that IEA remains very committed to helping its OECD member nations avoid supply shortfalls. For example, it produces reports such as *Energy Supply Security 2014: The Emergency Response of IEA Countries*, which catalogs and reviews the oil and natural gas emergency policies of IEA member countries (as well as countries such as China, India and other ASEAN members).[31]

One area in which the IEA has been particularly outspoken is in the need for greater worldwide investment in energy infrastructure. Its *World Energy Outlook*, released annually, repeatedly calls for this, seeing the lack of investment as potentially causing a supply shortage in the future. The 2014 edition stated that $48 trillion in investment—along with the policy planning necessary to make such investment possible—would be required between 2014 and 2040 to provide for all the expected energy needs during that timeframe.[32]

An analysis of the IEA in 2010 suggested that the agency is adapting to a growing and evolving global energy market.[33] In the decades since the 1973 petroleum supply issues, the IEA has focused its attention of "five structural developments of particular importance: the rise of new powers, climate change, peak oil, the concentration of oil and gas reserves, and the growing importance of new energy sources."[34]

While peak oil has been discussed for decades, the IEA has indicated a belief that petroleum production peaked in 2006 at 70 million barrels per day, although IEA reporting has varied on this issue and produced some criticism of the agency.[35] With the eventual decline in petroleum availability in mind, the IEA focuses significant attention on energy efficiency and reduced reliance on finite fossil energy sources. Renewable energy is actively and enthusiastically promoted by the IEA, which can address not only future energy shortfall, but environmental protection and climate change. Climate change research was in its infancy when the IEA was first formed in 1974. Since that time, the agency's focus reflects the consensus among its members that the effects of climate change brought on by massive carbon dioxide emissions from fossil energy use must be mitigated through reductions in fossil energy use.

The analysis of the IEA indicated that the agency has done a fairly good job of adapting to contemporary needs and realities, but has areas of weakness and unmet opportunities. The weakness of the agency is a continued focus on fossil energy and the promotion of energy reserves for supply stability. The analysis concludes that IEA should expand its energy "insurance" efforts beyond reserve management. The IEA could potentially have greater impact if it expanded to form direct ties to the BRIC nations—Brazil, Russia, India, and China—as their increased energy development and demands pose opportunities for the IEA to promote its goals in nations whose policies will likely have significant impact on climate change and energy security. Taken as a whole, OECD countries were once the largest consumers of energy in the world, representing greater than half of all energy consumed. Today, as China, India and other emerging economic powers are rapidly eclipsing OECD nations in terms of energy consumption, it is expected that OECD nations' energy consumption will represent less than 40 percent of energy consumed globally by 2030.[36] BRIC nations also pose significant opportunities for the further development of energy efficiency and renewable energy.[37]

The IEA emphases on energy supply and efficiency, renewable energy, and climate change are central to the promotion of energy security in the 21st century. An institution with a long presence in coordinating and informing energy policy in OECD countries offers a great deal of legitimacy to IEA efforts moving forward into a renewable energy future. The inclusion of BRIC and other emerging nations in the IEA mission could be highly successful and beneficial for at least two reasons. First of all, there can be a benefit to BRIC nations. Half a century ago, OECD nations were the majority consumers of fossil energy. Institutionally, OECD nations began to deal with many of the problems of dependence on imported energy supplies. In the late 1960s and early 1970s, OECD nations witnessed the rise of the environmental movement and concerted efforts to recognize and properly manage the pollution effects of fossil and nuclear energy use. Air and water quality issues became a prominent part of the environmental policy agenda. While the IEA is not a regulatory body, the agency has promoted critical energy and environmental quality policy analyses that readily inform domestic policy choices governing energy security and environmental quality. BRIC and other emerging nations could clearly benefit from the "lessons learned" by the IEA and its member nations as these rapidly emerging economies face increased pressure to deal with energy availability and energy use impacts in the form of air and water quality pollution.

Second, a more inclusive IEA that includes BRIC and other emerging economies in some capacity can also benefit OECD nations. In short, the developed world could learn a great deal from emerging world economies. Between 2008 and 2012, for instance, China increased its generation of electricity by renewables by over 68 percent.[38] Both China and Brazil are among the top ten nations in the production of biofuels as a clean fuel source. A coordinating body such as the IEA would likely benefit from including China, Brazil and India as more formal institutional members. This would allow for greater coordination and information sharing, while supporting efforts to developing better replication of success in developing policies and markets for biofuels and other alternative energy sources. At the same time, China, Brazil and other biofuel-producing nations—both OECD and non-OECD states—would mutually benefit from shared understanding of environmentally least harmful production methods using feed stocks less prone to environmental degradation.

OECD

The OECD was founded in 1961, emerging from its predecessor the Organization for European Economic Cooperation (OEEC), which was tasked in 1948 with rebuilding post-war Western Europe under the principles laid out in the Marshall Plan. To date, there are 34 member countries of the OECD, which is comprised of states that are the richest and most economically developed in the world.

The mission of the OECD is not centered primarily on energy. Its mission is broader than this, designed to promote collaborative approaches to promoting development, with an eye toward economic, social and environmental dimensions. It also seeks to develop shared standards and guidelines on areas as diverse as agriculture, taxation, and environmental protection. Energy security, however, does have a role in the OECD's work. Among its many energy-related activities, the organization is similar in its efforts to the IEA, and incorporates the work of the IEA into its own activities. It collects and publishes data on energy availability, production, use and trade. It seeks to coordinate policy among its members, particularly with regard to withstanding disruptions to energy supplies, and placing a strong emphasis on environmentally sustainable development. These efforts are directed in the service of energy security, which the organization defines in an expansive way, as it includes these many elements.[39]

A 2010 OECD report, entitled "Energy Security and Competition Policy," is instructive in this regard. It clearly elucidates several energy security issues that are important to OECD members, and puts the issue into sharp focus. Energy security is defined in this document as "vulnerability to disruption. Political turmoil, armed conflict, terrorism, piracy, natural disasters, nationalism, and geopolitical rivalry threaten, to varying degrees, to interrupt every trade in oil, natural gas, coal, and electricity."[40] A large portion of the report details efforts by the EU and member states in building purposeful redundancy into fossil energy and electricity supply chains so as to reduce the impact of any natural or human-caused disasters that could affect the supply of liquid fuels, natural gas, and electricity.

The report recognizes that there are different definitions of energy security in circulation, but points to a political economic model developed by energy policy scholars Douglas Bohi and Michael Toman in their book *The Economics of Energy Security*.[41] The Bohi and Toman approach focuses heavily on calculated levels of risk in supply disruption, on non-emergency pricing levels, and on the perception of risk. Given this understanding of energy security, it is not surprising to find that the OECD report identifies the following four key strategies for increasing energy security:

- Increased diversification of supply.
- Increased resilience through spare capacity and emergency stocks.
- Recognition of interdependence—that there is only one oil market and that the few regional gas markets may be melding into one.
- Timely information exchange, so that hoarding does not exacerbate shortages.[42]

In terms of diversification of energy source supplies, the OECD report identifies the use of wind and hydropower to make up for shortfalls in natural gas used in electricity generation. In addition, the report also mentions the use of nuclear power and coal as elements to be considered in the energy security enhancement work that lies ahead, in spite of their high levels of risk and public opposition that exists to their increased usage. The report also recommends cross-border sharing of electricity supplies. A shortfall in one nation facing limited natural gas supplies, for instance, could meet their respective electricity demands through better intra-continental power integration.[43]

The OECD report bows to recommendations for diversification of supplies made in the Green Paper drafted by the European Commission (EC) in 2006.[44] The EC and the OECD reports call for increased exploration and development of natural gas fields in North Africa, in the Caspian Sea region, and in the Middle East more generally. The intent is to develop a more diversified liquefied natural gas (LNG) supply chain capable of supplying the European market. While the reports do not specifically state it, the goal of diversifying the natural gas supply in Europe is related to past European-Russian supply dilemmas, such as the "abuse of dominance" illustrated by the Russian government in shutting off natural gas supplies multiple times as a result of repeated disputes between Russia and Ukraine.[45]

The OECD report also recommends that member nations pursue a policy of resilience. Resilience, as the term is used in the report, turns out to mean "excess capacity" in terms of energy supply. Accordingly, national policies have tended to pursue excess capacity in various forms. For example, industrial users have been encouraged to invest in high capacity back-up generators to meet peak demand. Resilience policies, however, have the capacity to reduce competitiveness if the benefits and costs of resilience are not carefully balanced.

The OECD paper recommends an awareness of interdependence of energy suppliers throughout the supply chain, particularly with regard to retail markets within countries and across borders.[46] This awareness of interdependence also prompts a policy recommendation in the report of pursuing long-term natural gas contracts.[47] Long-term contracts with gas suppliers represent a way of ensuring supply, and they act to encourage energy resource exploration by suppliers. Of course, without diversification, the problem with long-term contracts can be seen in the case of Russia's abuse of its market dominance. Long-term energy contracts in this case that are held by OECD member countries with Russian energy suppliers, put these countries, some of which are also members of NATO, at a major disadvantage in responding to Russian foreign policy.

OECD policy recommendations for global energy security in the 2010 report do not place a special emphasis on renewable energy sources. This same approach is apparent in other OECD reports, including its 2007 *Energy for Sustainable Development* report.[48] Instead, the OECD has taken a broader approach, pointing to potential gains in efficiency, along with policy measures directed toward regulation, R&D, transparency, investment, responsible business conduct, and especially market competition. The OECD has emphasized the need for multiple suppliers in the marketplace, both to make pricing competitive and to reduce overreliance on a small group of suppliers and few supply chains, though integration of supply and distribution networks can play an important role in increasing energy security.[49]

Energy Charter Treaty (ECT)

The Energy Charter Treaty (ECT) is a multilateral agreement that seeks to promote governance of global energy security issues. It was signed by 49 nations, including Russia and the European Union, in December 1994, though Russia never ratified the treaty and later withdrew in 2009. The treaty emerged in the aftermath of the collapse of the Soviet Union, with the aim of creating connections among EU countries and the Russian Federation. As Russia holds vast petroleum and natural gas reserves, properly accessing these reserves and transporting a range of energy products to markets worldwide required tremendous capital investment on the part of Western energy companies. Beyond the matter of the former Soviet Union and its vast energy reserves, there was a push to have an international agreement on non-discriminatory energy production and economic investment among major energy producing and consuming nations.[50]

As the Secretary General of the treaty stated upon its signing, the ECT provides "a multilateral framework for energy cooperation that is unique under international law … [it] is a legally-binding multilateral instrument, the only one of its kind dealing specifically with inter-governmental cooperation in the energy sector."[51] The treaty represents an effort to reduce the politicization and militarization of energy markets, and to help in this effort, the "strategic value of those rules is likely to increase in the context of efforts to build a legal foundation for global energy security, based on the principles of open, competitive markets and sustainable development."[52] The Energy Charter Treaty focuses on three major energy issues—namely, those of investment, trade and transit. One portion of the treaty promotes energy investment relationships among citizens, residents, and corporations headquartered in member nations. The treaty does not allow for discriminatory practices in the purchase or sale of energy concerns between ECT members. "The fundamental objective of the Energy Charter Treaty's provisions on investment issues is to ensure the creation of a 'level playing field' for energy sector investments through the Charter's constituency, with the aim of reducing to a minimum the non-commercial risks associated with energy-sector investments."[53] The trade provisions in ECT are grounded in the WTO principles of non-discrimination and the liberalization of trade.[54] And with respect to transit, the Energy Charter Conference (the governing unit of the ECT) has developed best practices templates for energy-producing nations in energy transportation, particularly pipeline development and non-discriminatory transportation practices.

The Energy Charter Conference is an open forum for ECT signatory nations and recognized observer nations to discuss issues related to energy-related investment, international trade and long-distance transit (the United States is a member in an observer role, it is not a signatory nation). As a governing unit, the Conference provides an opportunity for nations to reinforce their mutual pledge to one another in fair dealing in energy market transactions.[55] In spite of the nature of the ECT as a legal instrument, the Conference itself is not a venue for the arbitration of differences between nations or the corporations of individual stakeholders, and it possesses no enforcement authority. If a corporation or an individual entrepreneur in a member nation has a grievance against a member nation, the parties may choose to submit the dispute to arbitration, and an arbiter of the parties' choosing manages the arbitration:

> Foreign investors, individuals, or companies in dispute with a contracting state concerning an existing investment may choose to submit the dispute for arbitration. The investor may choose which of the available arbitration procedures to pursue, with or without the agreement of the host state. The provisions apply only to documented failures to honor obligations on the part of a host contracting state. The provisions do not apply to foreign investors that fail to fulfill their obligations. In such case, the host state must rely on either national domestic law or on the terms of the investment agreement.[56]

In other words, the Conference is designed only to hold states accountable, not private parties, and it serves more as a coordinating body than a directive entity. Moreover, it lacks the ability to enforce the terms of the treaty agreement.

The withdrawal of Russian Federation in 2009 brings into doubt the original goal of the treaty, which was to develop stronger institutional relationships among the energy sectors in Europe and Russia. Agreeing to the treaty obligations would have been likely to significantly weaken the country's control over the energy sector. One of the key items at issue has been the implementation of the "Third Energy Package," adopted by the EU in 2009, which has sought to clearly define market obligations among market participants. It requires, among other things, the unbundling of supply and transit of energy resources, and third party access to pipelines and transportation networks. Under the rules, no supply or production company (such as Gazprom) is permitted to hold a majority share of a transmission system (e.g., a pipeline). Alternatively, if a supply company were to own a transmission network, it would have to turn operations over to an independent system operator. These provisions have the potential to greatly impact Russian energy companies such as Gazprom and their operations in Europe. To that end, not only has Russia ended its affiliation with the ECT, which has sought to support and implement these rules, in 2015 Russia filed a dispute with the WTO, charging that the implementation of the Third Energy Package violates the WTO.

International Energy Forum

The International Energy Forum (IEF) was established in 1991. Unlike the OECD and the closely affiliated IEA, the IEF formally brings together OECD and BRIC countries as well as OPEC member states. The IEF serves as a coordinating body between energy-producing and energy-consuming nations. Its role has been to promote dialogue and transparency, as opposed to providing a forum for negotiations and decision making.[57] The permanent secretariat of the IEF is headquartered in Riyadh, Saudi Arabia. The IEF counts 88 member nations, which collectively account for approximately 90 percent of the petroleum consumed annually. The organization focuses attention on market transparency and the sharing of methodologies governing energy supply and demand forecasts. Forum dialogue is intended for member nations to share their assumptions and interests, as well as to discuss environmental concerns related to fossil energy. The IEF focuses primary attention on energy supply security as related to fossil energy, specifically petroleum and natural gas.[58]

A key policy component in IEF is the Joint Organizations Data Initiative, which is more commonly known as JODI. JODI is intended to provide updated global petroleum and natural gas data from over 90 oil- and gas-producing nations, thereby increasing market transparency. These data are intended to help nations to pursue energy policy strategies that best meet their energy needs given supply projections, and is intended to reduce uncertainty about energy supply. The energy data reported speaks mainly to current domestic production and demand, and importation of petroleum and natural gas. JODI does not, however, report known and estimated undiscovered petroleum and gas reserves, as many countries do not want this information to be made public, even though this reserve information would be particularly useful in energy security planning.

In the short term, the IEF and JODI provide critical data on petroleum and natural gas supplies and disposition as well as a forum to discuss energy goals of member states. In the long term, however, the IEF and JODI may prove less relevant to the promotion of energy security, as renewable and alternative energy sources play a more dominant role in meeting global energy needs. For now, however, the IEF is identified here as a part of the global energy security dialogue and likely will remain so as long as fossil energy plays a prominent part of global energy demand and supply.

The Limitations of Global Institutions

These four global institutions, along with others that address energy security from a multilateral platform, have important limitations on their capabilities. They have no direct control of energy resources; they cannot compel states to act differently when states' interests diverge from institutions' rules, norms and procedures; and, of course, they have no enforcement mechanisms. As Ann Florini and Benjamin Sovacool argue in "Who Governs Energy? The Challenges Facing Global Energy Governance," the shortcomings in global governance are numerous:

> Nationally, few if any governments are well structured to govern energy issues, much less participate in systems of global energy governance ... Internationally, the governance picture is even more incoherent. Energy is governed piecemeal, mostly in ad-hoc responses involving specific countries or groups of countries and any of a wide number of non-governmental actors.[59]

Nonetheless, these institutions do have important value. The collection and dissemination of data, along with the analysis of such data and salient issues, in numerous reports and databases, allows for greater transparency and accountability of governments, while identifying key challenges and threats that should be addressed. To that end, the IEA, OECD, ECT and IEF can be appropriately valued for their efforts to facilitate expanded energy markets, transparency, trade and investment, renewable energy, energy efficiency, and a reduction in greenhouse gas emissions, with varying emphasis on different economic, political and environmental goals. However, within this context, it is national interests and national policy on the part of sovereign states, and often times the interests and actions of non-state actors, that govern energy disposition and priorities. When interests diverge and/or when multilateral efforts fail to secure agreements and cooperation, the narrower interests of states and non-state actors clearly come to the fore, eclipsing the role and ability of international institutions in the pursuit of energy security. It is in this context that we turn to the consideration of a few key areas of concern for US energy policy and security.

Regional Issues and Concerns

Reducing the politicization and militarization of the global trade in energy as an overarching goal is a tall order, and one often derailed by the contingencies of global politics. Looking at US energy security concerns from the late 1960s and early 1970s, along with more recent developments such as the *Energy Policy Act of 2005* and the *Energy Security and Independence Act of 2007*, there has been great focus and attention on the strengthening domestic protections afforded by increased supplies—development of unconventional fossil, alternative, and renewable energy—that would buffer the nation against the global energy market and its vulnerabilities. Perhaps the conclusion to be drawn is that policy changes have worked as intended, albeit with many tradeoffs and costs. After decades of importing the majority of oil it uses, the US shifted the balance and began producing a majority of the petroleum it consumes on a daily basis. The domestic natural gas market, which in the early 2000s was set to increasingly rely on liquefied natural gas imports, is looking to export natural gas. Growth in renewable energy supplies has been exponential and the markets for solar and wind seem able to maintain continued expansion with no clear decline. Hybrid vehicles, advances in energy efficiency, regulation of carbon dioxide emissions, and smart grid technology also promise reductions in fossil energy consumption. This is not to say that the US has or will soon achieve complete energy security. Still, this evolution can be understood as a step forward on the path to a more depoliticized, demilitarized energy paradigm that reduces vulnerability, at home and abroad, to the disruptions of external forces.

These vulnerabilities come in a variety of forms, including the many conflicts in the Middle East that in general represent a threat to the production and transportation of oil, (thus impacting the goals of reliability, affordability and diversification); the growing demand for energy in Asia, (impacting the goals of abundance and affordability); and the assertiveness of Russia to politicize energy markets and provoke crisis in Eastern Europe, especially Ukraine (impacting national security concerns).

The issues involving US energy security vis-à-vis the Middle East have been well documented in numerous other places, and so this issue is not addressed extensively here, beyond acknowledging major US interests in the region. Taken as a whole, US interests in the region are numerous:

- Alliance with the State of Israel.
- Alliance as a co-member of NATO with Turkey.
- Significant past and present role in Iraq, including airstrikes against ISIS, the presence of several thousand troops in training and advising capacities, and what looks like a growing presence to further combat ISIS.
- Stopping the Iranian nuclear program and curbing the country's regional influence.
- Military basing agreements with Iraq, Bahrain, Kuwait, Qatar, and the United Arab Emirates.
- Friendly relations with oil-producing states.
- Open sea lanes.
- Stability and, if not peace, then an absence of war.
- And of course, ensuring the continued flow of oil.

Some of these interests are more related to energy security than others, but all of them have the potential to adversely affect energy security. After all, if there were no oil in the region, US interests and the US presence in the region would undoubtedly be far smaller. The region holds the largest proven petroleum reserves in the world, approximately 808 billion barrels of proven reserves as of 2015, led by Saudi Arabia (268.3 bbls), Iran (157 bbls), Iraq (144.2 bbls), Kuwait (104 bbls), United Arab Emirates (97.8 bbls), Qatar (25.2 bbls).[60] For this reason, the US has maintained longstanding alliances and business involvement with major petroleum producing nations in the region.

At present (2015), two of the most pressing issues revolve around the US interest in combatting the growing reach of ISIS (the Islamic State of Iraq and Syria), and checking the power and influence of Iran. ISIS currently controls significant portions of Western Iraq and portions of Eastern Syria. ISIS declares itself to be a world caliphate, and it proudly demonstrates its well-documented violence and brutality. The US accurately terms the organization a terrorist state. Under President Obama, US engagement in the fight against ISIS has involved the supply of weapons, food and equipment as well as targeted air support for coalition forces allied against ISIS. The United States has even provided air support to Iranian military forces operating with Iraq's willing approval in the fight against ISIS.[61] The US and other nations are also actively involved in providing signals interception and analysis, land-, air- and space-based intelligence to anti-ISIS coalition partners. The US has thus far resisted committing to a greater role, avoiding putting ground troops back into combat roles in Iraq, while encouraging regional powers to engage in combat against the group.

With respect to Iran, a great deal hinges on the agreement reached between the US and Iran in 2015 to curb Iran's nuclear program. Iran has rapidly expanded its nuclear program in the decade prior to the agreement, and has been listed by the United States as a state sponsor of terrorism for decades. Iranian-sponsored terrorism has often been directed against the State of Israel, a US ally, as well as other states friendly to the US, such as Saudi Arabia. In contrast to previous presidents, Barack Obama took the US relationship with Iran in a new direction, seeking to engage the Iranian government diplomatically to negotiate ways of reducing Iranian development of nuclear energy and the possible development of nuclear weapons. To the extent that the agreement is fully implemented and Iran is able to more fully develop its oil resources and access a larger segment of the global market, the impact will come to be felt with regard to oil supplies and prices.

China, India and the Pacific Rim

While the Middle East is generally characterized by rapidly moving crises, the Asian continent represents a longer-term, slower moving set of issues that can potentially impact US energy security. The continent is of enormous scale, and Asian nations such as China, India, and Japan are among the largest consumers of energy in the world. China is by far the biggest energy consumer in the region.[62] In 2013, China consumed 118.2 quadrillion British Thermal Units (BTUs) of energy, a figure which represents 54 percent of all energy consumed in a region extending from Japan to Afghanistan, and stretching from North Korea to Australia and New Zealand.[63] The second largest regional consumer of energy in 2013 was India, which consumed 26 quadrillion BTUs of energy, representing less than a quarter of China's consumption. Japan consumed 21.4 quadrillion BTUs in 2013. Combined, the three Asian "powerhouses" consumed roughly 165 of the 217 quadrillion BTUs used in the region; this represents approximately 76 percent of all energy consumed by over 55 percent of the world's population.[64]

In terms of petroleum, Chinese and Indian demand remains enormous and the nations have posted astounding rates of growth in petroleum consumption. While the US annual growth rate in consumption of fossil energy remained at an average year-to-year growth rate of 0.3 percent from 1980 to 2013, China and India averaged between five and six percent annual growth in the same period. China's daily consumption averaged in 2014 about 10 million barrels per day, compared to the US, which averages 19 million barrels per day. India uses approximately 3.5 million barrels per day. Sustained growth at these rates means that by 2030 China would consume over 23 million barrels of oil per day and India nearly 8 million barrels per day. Combined, the two nations are projected to consume more than the average daily production of petroleum from the Middle East (in the last ten years, the Middle East has produced approximately 26 million barrels of oil per day). Japanese petroleum consumption is likely to be less than 60 percent of India's daily consumption in 2030.[65]

The overall energy demands of China and India are rising at a rapid pace, and their anticipated energy demands would be unprecedented in human history. While China possesses domestic fossil energy sources, particularly coal, domestic oil supplies cannot even come close to covering the nation's massive energy needs. The result is expected to be even more intense global competition for energy resources.

Viewed through a lens of US energy security, this large and growing level of energy demand points toward potential problems, all related to one another: 1) to what extent might rising Asian demand for petroleum have an impact on the availability and price of oil for American consumers; 2) to what extent might competing, rivalrous demands for oil or natural gas prompt political conflict between the United States and China or other Asian nations; and 3) to what extent might Asian nations engage in political conflict among themselves, and could the United States be drawn into such conflicts? It appears that the first two questions posed above can be answered with cautiously optimistic answers of "a little bit, or not much ... yet." Though as Erica Downs has noted in her extensive work on China's energy security, it is not unreasonable to expect that over time, "Beijing is probably more willing to take actions to gain and maintain access to oil that run afoul of U.S. interests when those interests are not top U.S. foreign policy objectives."[66] With respect to the third question, it increasingly appears that in order to meet future energy demands, powerful Asian nations such as China, India, Indonesia, Malaysia and Japan are competing for resources, particularly with regard to upstream investment, rather than seeking ways to cooperate. It is reasonable to expect that the quest for energy security will allow for national self-interest to take precedence over regional cooperation, but this only points to the value of international institutions and agreements that can provide for equitable, fair procedures to ensure that countries will not benefit at the expense of others.

Private and publicly-owned business enterprises, such as CNOOC (China), Jindal Drilling Industries (India), and Japan Drilling Company (Japan) have invested heavily in the upstream development of energy resources in Africa, Asia, the Middle East, and Oceania.[67] Upstream investment establishes corporate property rights in the energy exploration, development, and transportation industries. A portion of the energy produced by the drilling concerns is competitively sold in the global energy market, while other portions for which a national oil company might hold an equity stake are transported to national markets. China, India and Japan benefit from the increased energy production to the extent that more energy products are available and increased supply is intended to translate into greater price affordability and enhanced price stability, critical aspects of energy security.

National energy companies in China, India, Indonesia and Malaysia play large visible roles in upstream energy investment and hold considerable energy industry assets. China National Petroleum Corporation has over $480 billion in total assets, while Petronas (a government-owned energy giant in Malaysia) and Indian National Oil Corporation have approximately $157 billion and $43.7 billion in assets, respectively.[68] Pertamina (a government-owned oil company in Indonesia) holds assets upwards of $50 billion.[69] In Japan, these upstream property rights are held by semi-private and private firms involved in the energy services industry.

Upstream investments in establishing energy security occur in two general, often overlapping forms, direct development or corporate acquisition. Direct development of energy resources involves either in-house development, mergers and acquisitions, or contract services. According to Jiang and Stinton, in a report entitled *Overseas Investments by China's National Oil Companies*, China's national oil companies spent over $18 billion in 2009 on mergers and acquisition of upstream international oil and gas companies, expanding their operations in 31 countries, the most significant being direct development in Kazakhstan, Sudan, Venezuela, and Angola.[70] For example, in Sudan and South Sudan, Chinese firms own a large stake in Sudanese and South Sudanese oil production operations, and 80% of South Sudan's oil goes to China.[71] In Iraq the China National Petroleum Company owns stakes in the Rumaila and Halfaya oil projects, and PetroChina has a 25 percent stake in Iraq's West Qurna 1 project.[72] Not unsurprisingly, Chinese investments, acquisitions and direct development can and have been viewed as a threat to Western energy security interests.[73]

While China's focus on energy security has been a more recent phenomenon, Japan has long had an interest in acquiring and developing upstream equity stakes in overseas petroleum and natural gas. Since the 1920s, Japan's demand for petroleum has outstripped its rate of production by a great deal. Early on, the national government saw energy security as a priority and established a national oil company as well as allowing a form of holding company that would strongly encourage upstream energy investment. In the early 1970s, British Petroleum approached the Industrial Bank of Japan with an offer to sell their stake in Abu Dhabi's offshore concession area, which was eventually sold to the Japan Oil Development Company (JODCO).[74] JODCO holds a 12 percent equity stake in two of the largest offshore oilfields in Abu Dhabi.[75] Japanese investment in overseas upstream energy development has not always been profitable, but the investments have been made with greater priority being given to the continual search for energy security in the form of fossil energy imports over short-term returns and profitability.[76]

While representing very different political systems, both Japan and China share some similarities in the way national energy security is promoted through national policy. Japan has used a mixed system of private, semi-private, and public finance, private banking, and upstream and downstream energy developments and mergers and acquisitions to promote its national energy security. China has engaged in very similar behaviors, but with a much larger and more evident role for the state and its priorities. India, however, provides a good example of how existing political, economic and social systems can exacerbate energy security vulnerabilities.

With respect to India, the country's energy consumption has grown rapidly, but it could potentially be encouraged to grow even faster. In a 2007 article entitled "Energizing the Indian Economy: Obstacles to Growth in the Indian Oil and Gas Sector and Strategies for Reform," Krishnan Devidoss points to a number of roadblocks to Indian energy development through direct investment and merger and acquisitions. "Unlike China, India has not achieved its potential because of its investment climate."[77] Despite being the fourth largest economy in the world, second largest in population, India's energy security is threatened in good part by internal challenges:

- corruption in the Indian bureaucracy;
- contractual problems, based on investor fear of abrogation;
- difficult access to credit, which inhibits privatization and mergers and acquisitions;
- onerous tax laws; and
- foreign policy and security issues related to opposition to a proposed pipeline with Iran.[78]

Devidoss points out that corruption is associated with lower levels of foreign investment. Fear of contracts not being properly honored adds separately to the risk in the business climate and may reduce investor interest in long-term, capital-intensive projects. Governments have been known to unilaterally cancel contracts with foreign energy investors on the basis of nationalistic interests. Credit costs in India are among the highest in Asia, which does not encourage investment. Additionally, in the area of mergers and acquisitions and in the realm of privatization, progress has been limited due to concerns over potential job losses. India has a strong socialist strain to its politics and reducing employment through privatization in an effort to increase efficiency and effectiveness tends to be highly unpopular among Indian voters. Tax liabilities facing energy exploration companies can be particularly burdensome, as India does not financially recognize the losses incurred by energy exploration that fails to discover energy supplies that are commercially viable. In other words, energy exploration companies face the added burden of taxes without the ability to write down losses incurred through their energy exploration activities. Finally, Devidoss addresses the international climate surrounding a proposed 1.6 billion cubic feet gas line running from Iran to India.[79] The United States raised an objection at the time due to its interest in economically weakening Iran due to the latter nation's nuclear enrichment program. The US dropped its objections to the pipeline, but the case does illustrate the role of international relations in shaping energy transportation via pipeline or sea channels.

Complicating energy security matters in Asia, beyond questions of energy demand and upstream investment to secure supplies, are growing tensions with regard to several islands whose ownership is in dispute, and which may hold large fossil energy reserves. Several countries have demonstrated increased interest in staking claims to the Spratly Islands, a chain of more than 100 uninhabited small islands and reefs whose ownership is disputed by China, Taiwan, Malaysia, the Philippines, and Vietnam. The islands are considered to be important largely for economic reasons. Currently the area is used for commercial fishing, but there are also potentially large—though unexplored and unproven—offshore reserves of oil and natural gas. Establishing sovereignty over the islands would afford the owner economic rights to the continental shelf and all the associated economic benefits. In support of these rival claims, China, Vietnam, Malaysia, and the Philippines have all sent small military forces to occupy some of the islands. All the disputants have pledged to resolve their differences peacefully, but each, particularly China, have taken steps to augment their presence and further establish their respective claims.

Another dispute, this one between China and Japan, involves rival claims upon a group of several uninhabited islands called the Senkaku in Japan and the Diaoyu in China. The islands are owned by Japan, which incorporated them in 1895 after a war with China, but China has increasingly asserted in recent years that the islands have been a part of its territory since ancient times, and that post-World War II settlements between the US and Japan did not properly address the rightful ownership of the islands. Again, like the disputed claims over the Spratly Islands, this issue involves the potential of exploiting what could be proven to be large fossil energy reserves.

Oil resources are one of the reasons attributed to rival claims on the Spratly and Senkaku/Diaoyu Islands, but there are additional strategic interests at stake that go beyond narrow, energy related concerns. Countries in the region maintain an interest in the area as a key supply route for goods and services in the South China Sea region. And China in particular has a defense interest in controlling these waters for open access to its growing Pacific fleet. Larger issues, however, revolve around the respective role that China can and will play in the region. For example, the dispute over the Senkaku/Diaoyu Islands prompted Japan in 2012 to fully assert its ownership by purchasing the islands in the chain that were still held privately. The government of China claimed that this was a provocation and sent surveillance ships and patrol aircraft into the area. Japan in turn scrambled fighter jets. In 2013 China declared the airspace over the island to be part of a new Air Defense Zone, requiring that "all aircraft intending to enter the zone had to file flight plans with the Chinese authorities, maintain radio communications and follow the instructions of Chinese controllers—or face 'defensive emergency measures.'"[80] While Japan's two major airlines complied with the notice, the United States made a point of refusing to recognize the claim and sent B-52s into the area without notifying the Chinese.

Disputed claims might be peacefully solved bilaterally or through multilateral forums, such as ASEAN, of which China and other disputant nations in the region are members. There are also other countries with an interest in these disputes, such as the United States and India. India's historical ties to the region and China's support for Pakistan have brought India into the South China Sea dispute as an interested party. India seems to view the situation as an opportunity to contain China's hegemonic ambitions. The United States has also become involved in the disputes, in part to protect its own historic influence in the Western Pacific as well as to maintain strong alliances with key regional actors such as Japan, South Korea, and Taiwan.[81] The United States has also strengthened ties with Vietnam, which sees the United States as an important player in helping Vietnam to thwart China's regional ambitions.

Russia, Europe, and Natural Gas

One of the world's largest energy producing and exporting nations, Russia is a key energy supplier to the European Union. Approximately one-third of the natural gas used in Europe comes from Russia, which is the EU's largest supplier of natural gas. The mostly state-owned energy giant Gazprom is the supplier, and it earns roughly $80 billion a year in revenues from the European market. This relationship leaves the EU highly vulnerable to Russian energy policy shifts, and is further complicated by the fact that over half of this natural gas is being transported in pipelines that pass through Ukraine (this figure has come down from 80% in 2009), with which Russia has maintained a longstanding dispute over prices, supplies, debts and transshipment.[82] With respect to the American energy security dilemma, the entire Russia/Ukraine/Europe natural gas conflict is bound up with larger issues involving Russia's relationship with the West, with former Soviet Republics, and with its status as a Great Power in global affairs.

Russia earns almost 70% of its foreign exchange from oil and gas exports, totaling $350 billion in 2013, and 50% of its federal budget comes from revenues from oil and gas.[83] This creates both opportunities and vulnerabilities for the country. It must continue to keep exports flowing or risk economic adversity, but it can also use the dependence of others to achieve political ends. Indeed, the growing ability of energy rich states to develop an assertive foreign policy—whereby national power is in part a function of being rich in energy resources—is reflected in the evolution of Russian's foreign policy.[84] Russia under Putin has asserted control over parts of Georgia and Moldova, and most notably it took Crimea from Ukraine in 2014 in a rapid military-political operation that surprised much of the world with its speed and audacity. It has since supported a separatist movement in eastern Ukraine. At the same time, Russia's relations with the EU, and especially the United States, have deteriorated over a number of issues (not only Ukraine and Crimea, but also missile defense, response to terrorism, economic cooperation, and Iran). The conflict regarding natural gas supplies is a microcosm of these larger issues, and exists as both cause and consequence of this worsening set of relationships.

Russian-Ukrainian conflict over natural gas supplies has been intermittent since the Ukraine became independent of the former Soviet Union in 1991. When it was part of the Soviet Union, Ukraine enjoyed plentiful supplies of cheap natural gas piped in from other parts of the country. This practice continued when Ukraine became independent, even as the relationship became one between two sovereign countries. In addition, the Soviet Union established a natural gas trade with Western Europe during the Cold War, and after the Cold War ended this economic relationship continued, supplying Russian gas via pipelines going through Ukraine.[85] In 2006, the ongoing issues of the price and quantity of Ukrainian gas purchases from Gazprom, along with outstanding Ukrainian debt to the company, came to a head, and it became clear to the world that this was far more than a business disagreement, but a political conflict.

The Russian-leaning government of Ukraine had been ousted in the "Orange Revolution" of 2005 and was replaced by a government more interested in expanding economic ties with the EU. The political and economic orientation of Ukraine—whether it would be tied more closely to Russia or move closer to the EU—came to be seen as emblematic of a larger question. It was an indicator of Russian power, economic vitality, and its ability to reestablish itself as a force to be reckoned with in its "near abroad" and around the world. Russia's response, which involved trying to weaken the new government, was to end the era of cheap, subsidized natural gas to Ukraine. Russia raised the price from $50 to $220 per thousand cubic meters, and demanded payment of outstanding debt, or it would cut off Ukraine's supply.[86] Failure to reach agreement resulted in a cutoff of a few days in January 2006 before a settlement was reached. An added element of the dispute is that Ukraine was accused of taking gas meant for transshipment to other European destinations and diverting it for domestic needs, something Ukraine later admitted doing. A similar, but longer gas cutoff occurred in January 2009 for the same reasons, in which a dispute over price, quantity, debt and diversion of supplies meant for other countries led to a natural gas cutoff. This time, several countries in Europe, not only Ukraine, were adversely impacted, as Russia flexed its muscles.

The latest iteration of the dispute between Ukraine and Russia—involving the takeover of Crimea and the Russian support for a separatist movement in eastern Ukraine—has raised the stakes not only of this confrontation, but with respect to Russia's foreign policy in general toward Europe and the United States. The violent ouster of a pro-Russia president in Ukraine who refused to sign a trade agreement with the EU as promised, Russia's subsequent invasion and annexation of Crimea, Ukraine's increasing ties to the West, the imposition of economic sanctions by the US and EU on Russia, and the rapid deterioration of Russia's relationships with the US and EU, have all far eclipsed the issue of natural gas prices and supplies.

By early 2015, the falling price of oil and (possibly) the imposition of economic sanctions had caused some degree of economic harm in Russia, as both revenues and the value of the ruble fell and sharply reduced Russia's foreign exchange earnings. Economic pressure has not caused President Putin to alter Russian foreign policy, and this is unlikely to change. Russia again cut exports to Europe via

Ukraine in the fall and winter of 2014–15, reducing exports by 60%, though the cutoff was not as severe as it could have been due to a mild winter in Europe and sufficient stockpiles of natural gas in several European countries. Part of this reduction involved Russia's cut of supplies to Slovakia, Poland and Hungary in retaliation for their shipments of gas to Ukraine. Russia cited technical problems and filling up storage facilities for the winter as the reasons for the reductions, but as Prime Minister of Slovakia Robert Fico stated, "gas has become a tool in a political fight."[87] Gazprom also continued to wrangle with Ukraine in 2015, raising the price of gas to the country from $268 to $485 per thousand cubic meters (the highest in Europe), demanding prepayment for some of this supply, as well as payment of outstanding debt, which Ukraine says it will only pay once the lower price is reinstalled.[88]

Though the particular details of this issue will change, perhaps rapidly, the challenges and larger set of issues remain the same. From the point of view of the United States in the pursuit of energy security, Russian behavior is particularly worrisome. While European countries seek to ensure natural gas supplies via markets that can remain as free of geopolitical conflict as possible, via instruments such as the Third Energy Package, Russian policy seems aimed to use Gazprom, a state-owned company, and energy supplies to consolidate control over the energy sector, and to use this control in the service of political gains, which may not always coincide with strictly commercial interests. To the extent that Russia can exploit its advantages as a major supplier of natural gas to Europe, this can further strengthen Russia, keep Europe dependent on Russian natural gas, cause a potential split in US and European interests, and leave the US with little ability to act to change these circumstances toward its own favor, at least in the short term.

The United States has not been held blameless in this matter, and has been criticized as exacerbating problems with Russia over several years. As John Mearsheimer argues in "Why the Ukraine Crisis is the West's Fault," the United States and its allies share responsibility for the crisis in Ukraine. "The taproot of the trouble is NATO enlargement, the central element of a larger strategy to move Ukraine out of Russia's orbit and integrate it into the West."[89] Indeed, Russia's regional concerns did become greater with the expansion of NATO to include many Eastern European nations formerly aligned with the Soviet Union. This was a break from the promise made by the United States as the Cold War came to an end. In 2008, the consideration of further NATO expansion to include Ukraine further fueled Russia's concerns and may very well have been a cause of the Russian natural gas shut off in January 2009. In addition, the United States has long maintained an effort to build a missile defense system, and for a time sought to station some of the system sites in Poland and the Czech Republic. President George W. Bush pushed forward with this goal in spite of the adverse impact on US-Russia relations, as Russia saw the system as being directed at them, despite US protests to the contrary. Even though President Obama later abandoned the plan to build a missile defense system in Europe,

Russia's increasing wariness of the US and the West can also be attributed, in part, to this. This view, representing a Realist approach to international relations, suggests that the West's embrace of a liberal democratic order centered around global institutions, rules and norms, along with open trade and investment, has ignored traditional power politics, which is more in line with Putin's view and Russian foreign policy.

The critique of US policy also extends to the US vilification of Vladimir Putin, and the assertion that he is personally responsible for the crisis in Ukraine. Stephen Cohen points out that this is not only a dangerous element of US policy, but that it has no basis. The demonization of Putin is simply a substitute for a policy, "an abdication of real analysis and rational policymaking."[90] This view, which remains in the minority in the United States, argues contrary to the vast majority of representations in the US and the West, that Russia under President Putin can legitimately be understood as reacting to US power and influence, to its expansion and proposed expansion of NATO into Eastern European, and to the encroachment of Western liberal democratic ideals into the Russian sphere of influence.[91]

Nonetheless, even had it been in the power of the United States and Europe to avoid the growing estrangement and hostility between Russia and the West, it is the case that the US and EU find themselves responding to Russia's foreign policy in Eastern Europe. Moreover, the US and EU are seeking to remain unified in policy toward Russia, such that European dependence on Russian natural gas does not create too much distance between the US and EU regarding their respective responses to Russia. This means that reducing European dependence on Russian natural gas, and on Russian supplies shipped via Ukraine, has become an important aspect of diminishing the energy security dilemma.

In the midst of the crisis over Crimea, there was much discussion in the United States about the possibility of US shale gas serving as a replacement for the EU's Russian gas supplies. Such an arrangement could be a perfect match.[92] Achieving this, however, is not a short-term solution. It will take several years of planning, construction and government approval of LNG facilities in the US, along with the development of contracts to ship the gas and sell it in European markets, which have to be ready and able to receive LNG imports of massive supplies of natural gas.

In spite of the inability of the US to serve as a natural gas supplier to Europe anytime soon, the EU has already made inroads in this direction, taking a multi-pronged approach to pursue energy security with respect to natural gas. One path is to pursue greater energy efficiency to reduce demand, along with the expansion of renewables to expand and diversify energy supply. Second, the EU is diversifying its natural gas suppliers, so it can rely less on Russia and on good Russian-Ukrainian relations. This includes keeping sufficient stockpiles to counter the impact of supply cuts, expanding the capacity to import LNG, establishing pipeline routes to bypass Ukraine, and establishing pipeline routes from non-Russian sources.

There are 21 LNG terminals in Europe, including a new terminal in Lithuania that opened in December 2014, with more planned for Poland, Italy, France and Croatia.[93] Since many existing LNG terminals in Europe are not utilized at their full capacity, this allows for additional increase in imports at existing facilities. In addition, new interconnection pipelines between Central and East European countries are allowing for greater ability to transport natural gas in a crisis.[94] This approach is diversifying the source of supplies, and it is projected that EU reliance on Russian natural gas will likely decline greatly over the next half century.[95] Still, natural gas imports via LNG terminals are likely to be expensive, as Europe will have to compete with the Asian market, in which prices are higher than those in Europe.

With regard to new pipeline routes, the Nord Stream pipelines, which are jointly owned by Gazprom, along with German, Dutch and French companies, came into service in 2011 and 2012, transporting gas from Russia through the Baltic Sea to Europe. A proposed "South Stream" pipeline was also to be constructed to run through the Black Sea to Bulgaria, but this was abandoned by Russia in early 2015. Russia has subsequently proposed a "Turkish Stream" to bring natural gas to southern Europe via Turkey, a project that has been publicly welcomed by Hungary and Greece.[96] While these projects do not reduce European reliance on Russia, they do bypass the Ukrainian route and avoid the potential problem of supply disruptions due to Russian-Ukrainian conflict. These cases reflect that fact that the goal of diminishing reliance on transshipment through Ukraine is one that Russia shares with the rest of Europe. Another set of proposed pipeline projects, across what is termed the Southern Gas Corridor, involves efforts to bypass Russia entirely. These projects are designed to bring natural gas to Europe from the Caspian Sea across several countries—Azerbaijan, Georgia, Turkey, Greece, Albania and Italy—via a series of pipelines.[97]

The relatively unified front that the US and EU have maintained, along with economic sanctions and the measures taken in Europe to counter dependence on Russian natural gas in the wake of Russia's actions, demonstrate clearly that the US and its allies see this challenge as far more than a dispute over energy supplies. Their actions also seem designed to further demonstrate to Vladimir Putin that he may be overplaying his hand. To the extent that Putin is using Russia's vast energy resources as a tool of great power politics, he may also be prompting a reaction that could ultimately weaken and isolate Russia, while causing it economic harm.

Walter Russell Mead's analysis of this situation reflects both this aim, and the tension between the Realist and liberal internationalist worldviews. He argues as a Realist on the one hand that Putin is "in it to win it."[98] Mead characterizes Russian foreign policy under Putin as a project that aims to "take advantage of the coming failures and catastrophes of what he believes to be the grandiose and unsustainable Western project in Europe." Moreover, Mead asserts that Putin "believes that the American commitment to Europe is so weak that the United

States will not react in a timely or effective fashion as Russia sets about the revision of the European order."[99] The West is so immersed in the liberal internationalist project of building market democracies and global institutions (he refers to Western elites as the Davoisie), that it is missing the point that Putin doesn't see the world in the same way. He's a Realist looking to assert Russian power and hegemony, and global institutions and rules hold little value for Russia. For example, a Western-centered view (one that is also critiqued by Mearsheimer and Cohen) looks at how, even beyond the immediate issues brought on by outright gas shut offs, Russia has rejected efforts on the part of the EU to reduce energy security conflict through international agreement and shared responsibility. For example, with respect to Russia's decision to withdraw from the Energy Charter Treaty, participation could potentially benefit Russia in bringing new resources for the maintenance of the Russian pipeline network, along with upstream investment in natural gas exploration and development. Russia has pushed for more downstream investment in European energy distribution firms, but this can be viewed as Russia seeking a one-way relationship, and one that the EU would not be likely to see as beneficial over the long term. The critique of this point, as Mead, Mearsheimer and Cohen might argue, is that it completely misses the point, by misunderstanding the Russian worldview and assuming that the attractiveness of the American view is obvious to all.

At the same time, Mead also argues from the liberal internationalist point of view, and in this sense is consistent with US policy. In "The Open Ukrainian Society and Its Enemies," Mead suggests that Russia's actions offer a renewed opportunity to support the broad "allied project of building a liberal world order."[100] An open, liberal, democratic Russian state remains the gold standard, and in spite of what Mead would suggest is a combination of Russian assertiveness and Western cluelessness to date, it is still possible to achieve this desirable outcome. As Mead states:

> Putin has challenged us to a contest to see whether the ideas and values of the West work better in the old Slavic heartlands of the Soviet Union than the mix of mafioso thuggery and nationalist hysteria emanating from the Kremlin these days ... [he] has quite unintentionally given the West a second chance to promote the construction of a genuinely democratic and prosperous Russia ... the odds are heavily against him. Our job isn't to respond to his military probes in the Donbas as much as it is to grasp the nature of our advantages and to bring the immense advantages of the West into play in ways that demonstrate to Russia that the path Putin has chosen is a historical dead end. We didn't beat the Soviet Union on the battlefield; we beat it by forcing the Soviets leadership to realize their utter inability to compete economically, technologically and ultimately militarily against the kind of open and dynamic society the western world built after World War Two ... If a united West can help Ukraine become a stable, prosperous and democratic

country, then not only will Putin's challenge to Ukraine ultimately fail, but he will be very hard put to hold onto power in Russia.[101]

These types of arguments may appear to stray from the question of US energy security and its enhancement through the maintenance of fair, open global markets for oil and natural gas. However, they are central concerns with respect to the objectives of US energy policy, which seeks to make energy available, affordable, and reliable, while allowing for the achievement of national security objectives.

Concluding Thoughts: Abundance or Scarcity?

A key question of whether political and military conflict over energy will grow or recede depends upon how well countries will be able to provide for their own energy security, however they define it. Taking a narrow view of energy security as "sufficient supplies at affordable prices," and looking, again narrowly, at oil and natural gas, these concerns point, first and foremost, to the availability of these resources. It may or may not be the case that the demand for energy will grow faster than supplies over time, and that a country's power and global standing seem to be, in part, a function of its control over energy supplies.[102] Whether or not these statements prove to be true, to the extent that states prepare for such eventualities in addressing potential resource insufficiencies, the goal of energy security seems likely to embody an element of "resource nationalism." As we have argued, state actions such as jockeying for access to the Arctic, making claims upon uninhabited islands, or bypassing market mechanisms to purchase oil in state-to-state contracts, suggest that the market-based system of energy trading is coming under stress. The result has been a great deal of state-led activity in seeking access to energy. The longer-term fear is armed conflict over the acquisition of energy.

At the same time, there has been remarkable growth in oil and gas production among major producers such as the US, Canada, and Russia, along with significant investments and new production of oil and gas in places like Brazil, Israel, China and other countries. One need only to briefly look at the newspapers and energy trade publications to see evidence of new discoveries of oil and gas, growing investments, and new production. These developments suggest that energy supplies are on the increase. If this is true, and energy supplies are able to keep up with rising demands, then sufficient supplies at affordable prices will be more likely to remain available, and markets for energy will be more likely to continue to function without state intervention. This development would suggest that concerns over resource insufficiency would diminish, and with it, the prospects for state-to-state conflict over access to energy resources.

These two countervailing trends suggest that the global system of energy production and use, along with the actions of states, is being defined simultaneously by the sense of a) increasing energy abundance as supplies grow, and a diminished likelihood of conflict over energy acquisition, and b) increasing energy scarcity as global demand outstrips supplies, and therefore greater potential for conflict over securing energy supplies. We have pointed to the rising demand for petroleum and natural gas in China and India, which will continue to be crucial to this question, as there is no sign that the rise in demand will taper off, leaving less powerful regional and global actors to scramble to meet their energy needs. And of course, beyond narrow energy concerns, crisis will also continue to affect these two trends and impact the elusive goal of energy security. While we identify conflict in the Middle East and Eastern Europe as sites of possibly explosive conflict, we are acutely aware that such prognostications are routinely eclipsed by other, unanticipated contingencies.

Notes

1 *BBC News*, "Russia Plants Flag Under North Pole," August 2, 2007, http://news. bbc.co.uk/2/hi/europe/6927395.stm (accessed March 4, 2015).
2 Doug Stokes and Sam Raphael, *Global Energy Security and American Hegemony*, Baltimore, MD: Johns Hopkins University Press, 2013, p. 2.
3 US Energy Information Administration, *International Energy Outlook 2013*.
4 Michael Klare, *The Race for What's Left: The Global Scramble for the World's Last Resources*, New York: Metropolitan Book, 2012; *The Daily Telegraph*, "Russia Claims North Pole with Arctic Flag Stunt," August 1, 2007, www.telegraph.co.uk/news/ worldnews/1559165/Russia-claims-North-Pole-with-Arctic-flag-stunt.html (accessed March 4, 2015).
5 Christopher Dickey, "The Oil Shield," *Foreign Policy*, March/April 2006; and Dickey, "The End of Iran's Oil Shield," *The Daily Beast*, November 12, 2013.
6 Steve A. Yetiv, *The Petroleum Triangle: Oil, Globalization and Terror*, Ithaca, NY: Cornell University Press, 2011.
7 US Department of the Treasury, Remarks of Under Secretary for Terrorism and Financial Intelligence David S. Cohen at the Carnegie Endowment for International Peace, "Attacking ISIL's Financial Foundation," October 23, 2014; "How Does ISIS Fund Its Reign of Terror?" *Newsweek*, November 6, 2014.
8 Nina Poussenkova, "The Global Expansion of Russia's Energy Giants," *Journal of International Affairs*, Spring/Summer 2010, p. 113.
9 Thomas Friedman, *Hot, Flat and Crowded: Why We Need a Green Revolution – And How it Can Renew America*, New York: Farrar, Straus, and Giroux, 2008, p. 110.
10 Andrei Chang, "China Expanding African Arms Deals," *UPI Asia.com*, January 26, 2009.
11 Wu Lei and Liu Xuejun, "China or the United States: Which Threatens Energy Security?" *OPEC Review*, 31(3), September 2007, p. 227.
12 Doug Stokes and Sam Raphael, *Global Energy Security and American Hegemony*, Baltimore, MD: Johns Hopkins University Press, 2013, p. 2.
13 EIA, "Petroleum and Other Liquids, Spot Prices," www.eia.gov/dnav/pet/pet_pri_ spt_s1_d.htm (accessed September 29, 2015).
14 Council on Foreign Relations, *National Security Consequences of U.S. Oil Dependency*, Independent Task Force Report No. 58, 2006.
15 Council on Foreign Relations, pp. 19–23, 27.
16 Barack Obama, "Remarks by the President on the Economy—Northwestern University," The White House Office of the Press Secretary, October 2, 2014.

17 "Oil's Comeback Gives U.S. Global Leverage," *New York Times*, October 8, 2014, F7.

18 "Oil's Comeback ..." ; "An Oil Boom is a Power Boon," *US News and World Report*, December 1, 2014.

19 *YCharts*, "EU Natural Gas Import Price, and Japan Liquefied Natural Gas Import Price," https://ycharts.com/indicators/ (accessed October 9, 2015).

20 Federal Energy Regulatory Commission, LNG, "Existing and Proposed Terminals," www.ferc.gov/industries/gas/indus-act/lng.asp (accessed October 9, 2015); Doug McAdam and Hilary Boudet, *Putting Social Movements in Their Place: Explaining Opposition to Energy Projects in the United States 2000–2005*, Cambridge: Cambridge University Press, 2012.

21 Andreas Goldthau and Jan Martin White, "The Role of Rules and Institutions in Global Energy: An Introduction," in Goldthau and White, eds, *Global Energy Governance: The New Rules of the Game*, Berlin: Global Public Policy Institute, 2010, p. 5.

22 John Deutch, "The Good News about Gas," *Foreign Affairs*, Jan./Feb. 2011, p. 90.

23 Center for a New American Security, *Energy Rush: Shale Production and U.S. National Security*, February 2014, p. 6.

24 G. John Ikenberry, "The Future of the Liberal World Order," *Foreign Affairs*, May/June 2011, p. 61.

25 See, for example, Ann Florini and Benjamin Sovacool, "Who Governs Energy? The Challenges Facing Global Energy Governance," *Energy Policy*, 37, 2009, pp. 5239–48; Goldthau and White, eds, *Global Energy Governance: The New Rules of the Game*; Thijs Van de Graaf and Kirsten Westphal, "The G8 and G20 as Global Steering Committees for Energy: Opportunities and Constraints," *Global Policy*, Special Issue, September 2011, pp. 19–30.

26 Goldthau and White, "The Role of Rules and Institutions ...," p. 2.

27 Dimo Bohme, in *EU-Russian Energy Relations: What Chance for Solutions? A Focus on the Natural Gas Sector*, Potsdam: University of Potsdam, 2011, p. 285.

28 Van de Graaf and Westphal, "The G8 and G20 ...," p. 20.

29 International Energy Agency (IEA), "What We Do," www.iea.org/aboutus/ (accessed April 2, 2015).

30 IEA, "Member Countries," www.iea.org/countries/membercountries/ (accessed April 2, 2015).

31 IEA, *Energy Supply Security 2014: The Emergency Response of IEA Countries*, 2014, www.oecd.org/publications/energy-supply-security-2014-9789264218420-en.htm (accessed April 2, 2015).

32 IEA, *World Energy Outlook*, 2014.

33 Thijs Van de Graff, "Obsolete or Resurgent? The International Energy Agency in a Changing Global Landscape," *Energy Policy*, 48, 2012, p. 234.

34 de Graff, p. 234.

35 Richard G. Miller, "Future Oil Supply: The Changing Stance of the International Energy Agency," *Energy Policy*, 39, 2011, pp. 1569–1574.

36 de Graff, p. 235.

37 de Graff, pp. 235–40.

38 US Energy Information Administration (EIA), "International Energy Statistics," www.eia.gov/cfapps/ipdbproject/IEDIndex3.cfm?tid=79&pid=79&aid=1 (accessed March 4, 2015).

39 Organization for Economic Cooperation and Development (OECD), *OECD Contribution to the United Nations Commission on Sustainable Development: Energy for Sustainable Development*, 2007, pp. 9–11.

40 OECD, "Energy Security and Competition Policy," *OECD Journal Competition Law and Policy*, 11(1), 2010, p. 17.

41 Douglas R. Bohi and Michael A. Toman, *The Economics of Energy Security*, New York: Springer, 1996.

42 OECD 2010, p. 27.

43 OECD 2010, p. 28.

44 European Commission, *Green Paper: A European Strategy for Sustainable, Competitive and Secure Energy*, 2006.

45 OECD 2010, p. 37.

46 OECD 2010, p. 33.

47 OECD 2010, pp. 34–37.

48 OECD 2007.

49 OECD 2010, pp. 34–37.

50 Philip Andrews-Speed, "The Politics of Petroleum and the Energy Charter Treaty as an Effective Investment Regime," *Journal of Energy Finance and Development*, 4, 1999, p. 117.

51 Ria Kemper, "Foreword," *The Energy Charter Treaty and Related Documents: A Legal Framework for Energy Cooperation*, Energy Charter Secretariat, 2004, p. 13.

52 Kemper, p. 13.

53 Kemper, p. 14.

54 Kemper, p. 15.

55 Andrews-Speed, p. 117.

56 Andrews-Speed, p. 121.

57 Goldthau and White, "The Role of Rules and Institutions in Global Energy...," p. 8.

58 International Energy Forum, "Global Energy Security through Dialogue," *International Energy Forum*, 2012, p. 4, www.ief.org/_resources/files/latest-files/latest-ief-brochure.pdf (accessed March 4, 2015).

59 Florini and Sovacool, "Who Governs Energy? ..." p. 5239.

60 EIA, *International Energy Statistics, Petroleum Reserves*, 2015.

61 *The Washington Post*, "The US and Iran Are Aligned in Iraq Against the Islamic State—For Now," December 27, 2014; and *The Daily Beast*, "US Backs Iran with Airstrikes Against ISIS," March 25, 2015.

62 EIA, *International Energy Statistics, Petroleum Consumption*, 2015.

63 EIA, *International Energy Outlook 2013*.

64 EIA, *International Energy Outlook 2013*.

65 EIA, International Energy Statistics, Petroleum Consumption, 2015.

66 Erica Downs, "China," The Brookings Foreign Policy Studies Energy Security Series, Brookings Institution, 2006, p. 2.

67 Energy Business Review, "Japan Petroleum Exploration Co., Ltd., 2014," www.energy-business-review.com/companies/japan_petroleum_exploration_co_ltd (accessed March 11, 2014); Julie Jiang and Jonathan Sinton, *Overseas Investments by Chinese National Oil Companies*, International Energy Agency, 2011; and US EIA, "Countries: Japan," www.eia.gov/countries/analysisbriefs/Japan/japan.pdf (accessed March 11, 2014).

68 Fortune, "Global 500: Our Annual Ranking of the World's Largest Corporations," *Fortune Magazine*, 2012, http://money.cnn.com/magazines/fortune/global500/2012/snapshots/10939.html (accessed March 11, 2014).

69 Pertamina Annual Report 2014, pp. 10–11.

70 Jiang and Stinton, p. 7.

71 Phillip Manyok, "Oil and Darfur's Blood: China's Thirst for Sudan's Oil," http://api.ning.com/files/imbe9KBD0dNFdVFA8jUUWh9KpTWt-o5kytTNtCPt4SHHIVCzOriEEJ7PBySbr★fKcnZ3cmMXq3pH3PXdzLd38D8raS36R6eW/ChinaThirstforSudanOil.pdf; and "Even China Has Second Thoughts on South Sudan after Violence," *Los Angeles Times*, February 20, 2014, www.latimes.com/world/asia/la-fg-south-sudan-economy-20140220,0,3253847.story#axzz2ybCnWQ3j (accessed March 20, 2015).

72 "China Doubles Down on Iraqi Oil Gamble," *The Diplomat*, October 18, 2013, http://thediplomat.com/2013/10/china-doubles-down-on-iraqi-oil-gamble/, and "China Pledges to Pump More Funds into Iraq's Oil Sector, Infrastructure," *South China Morning Post*, February 24, 2014, www.scmp.com/news/china/article/143402 5/china-pledges-pump-more-funds-iraqs-oil-sector-infrastructure (accessed March 20, 2015).

73 Wensheng Cao and Christoph Bluth, "Challenges and Countermeasures of China's Energy Security," *Energy Policy*, 53, 2013, pp. 381–388; see also NBR, "China's Intentions for Russian and Central Asian Oil and Gas," *National Business Review* 9, 1998, p. 5; and Zhong Xiang Zhang, "The Overseas Acquisitions and Equity Oil Share of Chinese National Oil Companies: A Threat to the West but a Boost to China's Energy Security," *Energy Policy*, 48, 2012, pp. 698–701.

74 Gerald Pollio and Koichi Uchida, "Management Background, Corporate Governance and Industrial Restructuring: The Japanese Upstream Petroleum Industry," *Energy Policy* 27, 1999, pp. 813–832.

75 JODCO, "The Concession," Japanese Oil Development Company, Ltd., 2014, www.jodco.co.jp/english/business.html (accessed March 20, 2014).

76 Pollio and Uchida, 1999.

77 Krishnan A. Devidoss, "Note: Energizing the Indian Economy: Obstacles to Growth in the Indian Oil and Gas Sector and Strategies for Reform," *Boston College International and Comparative Law Review*, 30, 2007, p. 202.

78 Devidoss, p. 204.

79 Devidoss, p. 208.

80 "Regional Turbulence," *The Economist*, November 30, 2013.

81 Leszek Buszynski, "The South China Sea: Oil, Maritime Claims, and U.S.-China Strategic Rivalry," *The Washington Quarterly*, 35(2), 2012, pp. 139–156.

82 Frank Umbach, "Russian-Ukrainian-EU Gas Conflict: Who Stands to Lose the Most?" *NATO Review*, May 9, 2014, www.nato.int/docu/review/2014/nato-energy-security-running-on-empty/Ukrainian-conflict-Russia-annexation-of-Crimea/EN/index.htm (accessed April 19, 2015).

83 EIA, "Oil and natural gas sales accounted for 68% of Russia's total export revenues in 2013," *Today in Energy*, July 23, 2014.

84 Michael Klare, *Rising Powers, Shrinking Planet*, New York: Metropolitan Books, 2008; and Thomas Friedman, "The First Law of Petropolitics," *Foreign Policy*, May/June 2006, pp. 28–36.

85 Per Hogselius, *Red Gas: Russia and the Origins of European Energy Dependence*, London: Palgrave Macmillan, 2013.

86 Andrew Kramer, "Russia Cuts off Gas to Ukraine in Cost Dispute," *The New York Times*, January 2, 2006.

87 *Reuters*, "EU-Russia Gas Duel Deepens with Slovakia Supply Cut," October 1, 2014, http://uk.reuters.com/article/2014/10/01/ukraine-crisis-gas-idUKL6N0RW3TZ20 141001 (accessed April 19, 2015).

88 Umbach, 2014.

89 John J. Mearsheimer, "Why the Ukraine Crisis is the West's Fault," *Foreign Affairs*, September/October, 2014, p. 77.

90 Stephen Cohen, "Patriotic Heresy: Neo-McCarthyites Have Stifled Democratic Debate on Russia and Ukraine," *The Nation*, September 15, 2014.

91 Stephen Cohen, "Distorting Russia," *The Nation*, March 3, 2014.

92 Christoper Goncalves and Anthony Melling, "A Perfect Match," *Natural Gas and Electricity*, 30(8), March 2014.

93 Global LNG Info, "World's LNG Liquification Plants and Regasification Terminals," April 2015, www.globallnginfo.com/world%20lng%20plants%20&%20terminals.pdf (accessed April 19, 2015).

94   Umbach, 2014.
95   Sergey Paltsev, "Scenarios for Russia's natural gas exports to 2050," *Energy Economics*, 42, 2014, pp. 262–270.
96   *The Economist*, "A Twist in the Pipeline," April 14, 2015.
97   Trans-Adriatic Pipeline, "Southern Gas Corridor," www.tap-ag.com/the-pipeline/the-big-picture/southern-gas-corridor (accessed April 19, 2015).
98   Walter Russell Mead, "In It to Win It," *The American Interest*, January 27, 2015, www.the-american-interest.com/2015/01/27/in-it-to-win-it/ (accessed April 19, 2015).
99   Mead, "In It to Win It."
100 Walter Russell Mead, "The Open Ukrainian Society and Its Enemies," *The American Interest*, February 14, 2015, www.the-american-interest.com/2015/02/14/the-open-ukrainian-society-and-its-enemies/ (accessed April 19, 2015).
101 Mead, "The Open Ukrainian ..."
102 Klare, *Rising Powers, Shrinking Planet*.

7

CONCLUSION

Energy security is a complex issue, neither easily defined nor easily managed. Without ready access to continually available, reasonably priced energy sources much of the developed world would rapidly slip back in time in terms of commercial activity, economic dynamism and overall quality of life. In the modern era, people in industrial societies have largely divested themselves of many of the practices, skills, and tools associated with providing for themselves, eschewing traditional methods of living that make survival possible with limited energy resources. Traditional societies where household implements and modes of living require limited resources still exist, but they are now fewer in number, and are virtually everywhere rapidly adopting the relative comforts of industrial and post-industrial life wherever possible. In other words, globalization—fueled by massive changes in communications and transportation, along with trade liberalization and the rapid movement of capital investment—has made it increasingly possible for individuals and societies to rely on others to provide the goods and services they want and need. Greater numbers of people, firms, and countries rely on energy supply chains that extend around the world, with energy companies engaging in multinational partnerships to pursue projects, financed by investors from virtually all parts of the globe.

Globalism makes the evolution toward modernity of societies occur more rapidly, and often in ways that lead to greater forms of inequality and dysfunction as traditional societies experience the loss of their traditional survival skills and the replacement of basic tools with energy-hungry machines of all kinds. Young people rapidly leave agrarian communities to take up residence in cities to work in factories and myriad shops and stores associated with urban population centers. Individuals, firms and countries embrace the opportunities offered by globalization; this is indeed a reasonable expectation. Still, all of these changes contribute to problems of energy security, as patterns of energy use change in terms of the amount and type of energy demanded, and change as well in terms of the geographic origins and endpoints of a supply chain that makes such societal changes possible. The worldwide system of supplying, producing and consuming vast amounts of energy is remarkable in its complexity and capability, but it is also—due to this complexity—remarkably fragile and vulnerable to disruption. It is, in short, insecure.

Technology, markets and public policy continually interact to address the opportunities and challenges of energy security. Traditionally, public policy analysis involves the defining of a problem followed by the development of alternative strategies to be followed to solve the problem. Admittedly, most policy problems are more complex than the political process makes them out to be. Rival interest groups generally seek to take advantage of public concerns associated with a problem by offering their own narrow framing of the problem and formulation of remedies. In the case of energy security, however, the preceding chapters of this book have highlighted how there are a multitude of interconnected and continually evolving problems that do not all easily fit under one policy tent. Workable and lasting solutions, therefore, are highly unlikely to emerge for a very long time—indeed, if ever. In an issue area as fraught with peril as energy security, it would be irresponsible to suggest otherwise. At the same time, public policy is an integral element in the effort to achieve greater energy security. It has not been and cannot be otherwise. So, while policy (and the process of establishing and implementing policy) can be problematic in identifying problems and offering solutions, it is also indispensable. The goals of energy security are many—abundance, reliability, affordability, cleanliness and sustainability, and diversity, all without threatening national security or other major interests. Moreover, the range and magnitude of crises, simmering problems, and long-term vulnerabilities seem to have proliferated. All of this demands attention from governments.

Energy demand domestically and globally is both accelerating and changing in character at the same time. By mid-century, world energy demand is projected to increase by at least 50 percent. With that high rate of acceleration and widespread change comes the increased possibility of a monumental energy security failure. The lead time for accommodating shifts in energy resource types and figuring out reliable transportation to market has grown shorter due to the overwhelming and growing demand evident throughout the world.

Beyond questions of immediate demand, there are also a whole host of issues related to the intended and unintended consequences of the energy being produced and used. As we have argued, an energy security dilemma is an inherent part of the domestic and global energy system. The marginal social costs of using fossil energy are extremely high, particularly if one considers the prospect of global climate change and the occurrence of its associated adverse consequences. These societal costs are not "marginal" at all—to the contrary, the changes predicted by scientists may force changes to societies around the world and lead to severe impacts for the world's plant and animal species. Extreme temperatures are more likely to occur, with some regions of the planet experiencing extensive and prolonged drought, while other areas may witness long-lasting and heavy rain or snowfalls. Climatologists predict that there will be a dramatic increase in devastating weather events such as typhoons and hurricanes, as well as the permanent flooding of shoreline areas brought on by rising ocean levels. In short, two complex systems—the global environment and the global economy—are colliding with one another, with great unpredictability about the consequences on either system as this interaction deepens. Therefore, globalism in economic and commercial activity, along with environmental change in our shared occupancy of the planet, bring us closer together and demand a concerted effort to enhance energy security.

Another major change looms on the horizon as well, and while its exact timing remains unknown, it will most certainly occur at some point. The eventual depletion of fossil energy is a certainty, not to a point where none remains, but to a point where it is no longer economically justified to use it. In other words, the last barrel of oil will never be pumped out of the ground because it will cost too much to get it. This circumstance, at least with respect to petroleum, seems likely to occur sometime in this century, though it is important to note that the demise of fossil energy has been inaccurately predicted since the late-19th and early-20th century. Consequently, we will not venture to place any bets on the date and time of this particular change. The rate of energy use is increasing rapidly—the thirst for energy in the developing world continues to grow and low energy-use areas of the world are quickly becoming high energy-use areas. It follows that unless alternative energy sources can be more rapidly developed and subsequently introduced into widespread use to meet growing demand, energy security, in terms of abundance, reliability and affordability, will be threatened.

Renewable energy offers the possibility of diminishing the environmental impacts of our energy use. Federal and state policies in the United States provide strong support for renewables, through tax incentives, renewable portfolio standards, and spending public dollars on technology research and development. The Obama Administration has shown itself to be eager to promote environmentally oriented energy security plans. It adopted new rules for limiting carbon dioxide emissions and other pollutants from both new and existing power plants. If fully implemented (and successful in getting past various legislative and legal challenges), these rules could shutter the vast majority of coal-fired electricity generators in the country, to have them replaced by renewable sources in the form of solar, wind, and geothermal energy (though it is also expected that coal's replacement will, to a large extent, come from natural gas). Hydropower will remain an important renewable energy source far into the future, particularly in the developing world, but it will not be expanded in the United States due to its adverse impact on native fish species (the listing of several species of salmon under the Endangered Species Act, along with Oregon and Washington developing salmon recovery plans in the 1990s and early 2000s, signaled a fundamental change in the use of large hydropower facilities in the United States). Such efforts promote sustainability while advancing energy security goals related to supply, cost, reliability, diversification, and environmental health and protection. Still, public policy support for renewable energy here in the United States is not as strong as it is in "greener" nations such as Germany. The United States has not passed comprehensive national legislation to combat greenhouse gas emissions; it has not passed a nationwide renewable portfolio standard; its gasoline taxes are relatively low compared to most other rich, industrial nations; it has even allowed the Production Tax Credit, which has prompted a rapid expansion of wind power, to expire, and may allow the same thing to happen to the Investment Tax Credit.

There are many reasons for a less accelerated path to energy security through renewables than might be possible, all of which reflect some short-term challenges in the United States. Technical, financial and political challenges to rapidly scaling up renewable usage abound, as we have discussed in earlier chapters, but there is perhaps first and foremost the unanticipated impact of large-scale shale oil and natural gas development in North America that has occurred. As a direct consequence of major breakthroughs in hydraulic fracturing, the United States ranks as the largest producer of petroleum and natural gas in the entire world.[1] While the US government has heavily committed federal government budget resources to renewable energy source research and the commercialization of technologies, the fossil energy boom propelled by the use of hydraulic fracturing has often overshadowed the focus on renewable energy. Added to this development has been an ongoing change in the automobile industry, which has, as a result of achieving more aggressive CAFE standards mandated by federal regulation, substantially increased the fuel efficiency of the foreign and domestic vehicles sold in the United States.

The result from these developments has been affordable fossil energy prices for transportation, along with more manageable demand levels. The majority of this demand is being met with US domestic petroleum supply, and much of the remaining supply is provided by reliable North American trading partners. Considering that the newly developed petroleum and natural gas resources have played a large role in boosting the US economy, that public policy with regard to fracking has largely involved enabling its rapid deployment, and that Big Oil and numerous smaller energy companies in the United States have been virtually untouchable politically, it is not surprising to expect that the fossil energy revolution in the United States will continue to be a key defining element in the American energy profile for quite some time.

Energy security, however, even when viewed primarily in terms of a domestic focus, must be considered in a global context. Almost the entire world is increasingly hungry for energy. That same world is also a rather dangerous place, with many challenges that will likely have an unfavorable impact on energy availability, on safe and timely transportability, and on consistency in supply. Conflict between the Ukraine and Russia, conflict in Iraq and Syria, terrorist activity in multiple resource-rich countries, and Chinese claims on petroleum resources in Southeast Asia collectively represent heightened threats to energy security. Should new or expanded military conflict erupt in any of these places, it would likely result in noteworthy oil supply shortfalls and price spikes, and would possibly prompt new public policy efforts to promote domestic energy security.

The American Energy Profile, Today and into the Future

While global threats most certainly do exist, the current energy profile of the United States with respect to the mix of fossil energy and renewable energy can help to mitigate some of the risks associated with international energy supplies. Fossil energy abundance in our country and the ability to build and replenish a strategic reserve reduce the severity of shortfalls and price spikes. And over the longer term, growing use of renewable energy alternatives, such as biofuels and electric hybrid vehicles recharged using green energy, can and will further reduce the risks to American (and global) energy security.

Arguably, the nation's current fossil fuel-rich energy profile has proved to be highly useful to the United States as it has been engaged in its effort to manage the web of American involvements in the Middle East and Central Asia. Clearly, ISIS and other regional groups seek to weaken or destroy established nation states and redraw the map of the Middle East. Of note, the nations of the Middle East are not only the suppliers of much of the world's petroleum, but these nations are also crisscrossed with pipelines that transport oil and natural gas energy supplies bound for Europe and Asia. Capable, reliable and stable governments are needed in the region to resolve disputes, to quell violence, to maintain supply, to diminish price volatility, and perhaps importantly to ensure that contracts negotiated are honored. Such negotiated agreements serve as the very foundation of the international energy market.

Despite its current abundance of fossil energy riches, most American energy companies operate internationally, and the costs of doing business on an international scale are factored into prices set by international markets. In other words, the United States is not an energy-independent island in a sea of turmoil and rising conflict. One way or another, we suspect that global energy security needs will necessitate an ongoing active American role in safeguarding nearly all major aspects of global energy markets. The United States has assumed a key actor role in this arena over time, and that role will remain a central aspect of US foreign policy for the foreseeable future.

Regardless of the complexities involved, it is nonetheless the case that some aspects of energy security can be dealt with effectively through public policy. In the worst of cases, governmental action might come in the form of war or more limited military engagements. However, the most likely approaches to be expected on the part of the United States in carrying out its possible roles are more limited. One way of reducing threats to US and global energy security is to engage actively with countries such as China and Brazil (both of which are rapidly growing in terms of energy production and/or consumption) in realms directly connected to energy, and more broadly across the commercial, cultural and communication landscape of international relations. Cooperative ventures that highlight shared responsibilities and common goals may reduce the likelihood of open conflict for energy resources. For example, the US Export-Import Bank in 2009 made a preliminary commitment to Petrobras, Brazil's national oil company, to develop offshore oil fields accessing resources several miles below the sea in the "pre-salt layer."[2] In addition, American energy companies may in the future play a major role in Brazil's energy sector (they have not yet been major investors in developing these newly found resources). In the private sector more generally, there is likely to be an enlarged role for energy contractors who are capable of maximizing energy supply from known deposits through innovative drilling techniques. Relatedly, an increased role is likely to be given to energy facility and infrastructure security experts—working in cooperation with domestic and international military and intelligence services—providing security for drilling, refining, and transportation systems and limiting risks to security resulting from the actions of non-governmental actors. In a non-energy related area, the United States invited the Chinese Navy to participate in Rim-of-the-Pacific (RIMPAC) naval exercises to promote United States-China cooperation on security issues.[3] Given Chinese claims to offshore oilfields near Vietnam, Japan and other nations, the joint RIMPAC exercises are indeed timely. Cooperative security efforts of this type are sometimes effective methods of reducing miscommunication and forestalling conflict.

Domestically, energy security can also be enhanced to an important extent by reducing commercial and household energy consumption and improving the efficiency of electricity transmission. The Rocky Mountain Institute's publication *Reinventing Fire* offers a multitude of ways to invest in the nation's electric power infrastructure to make it more efficient, smarter, less prone to disruption and outages, and more cost-effective.[4] Public policy support, in the form of renewable portfolio standards, smart grid investments, a greenhouse gas cap-and-trade system, new technology development, and outright subsidization can all contribute to the energy efficiency promotion effort. In the transportation sector, commuting to and from work using personal automobiles creates a high demand for motor fuel and motor oils. Additionally, commercial truck fleets transport products from ports or from domestic manufacturing facilities to retail stores nationwide. When not transported by pipeline, energy products such as crude petroleum, refined gasoline and diesel, and natural gas are transported by commercial truck, in the former case to refineries, and in the latter cases to filling stations in every city and town in the United States. In order to get the gallon of gasoline a person might purchase at a retail station, a great deal of energy is consumed in the transportation process. Mandating more efficient vehicles, through increasingly stringent CAFE standards, might be one method of reducing fuel consumption for transportation. Other possibilities involve taxation and reg-ulation (admittedly unpopular and unlikely policies) that lead to the increased cost for personal vehicle transportation, such as gasoline taxes, carbon taxes, or feebates (imposing a fee for less-efficient vehicles while providing a rebate for more effi-cient ones). If the cost of driving a personal car increased, such changes in incentives might lead to a decline in personal vehicle use and a subsequent increase in either mass transit ridership or telecommuting. Reduced energy demand would reduce the pressure on energy markets in terms of supply, while reducing the impact and exposure to energy security risks, such as price spikes, temporary fuel shortages, and other unplanned disruptions in supply.

Though other analyses of energy security focus little on the subject of prop-erty rights, we argue that this realm of public policy is also central to domestic energy security. From the late 19th century and continuing forward to today, energy policy has focused principally on the maintenance of a ready supply of energy from domestic sources, largely involving fossil energy resource extraction from both private and public lands. Sub-surface resources, such as certain grades of petro-leum and natural gas, have been subject to the law of capture, which leads to energy "bonanzas" in regions where oil and gas have been discovered and exploited. As markets have matured and evolved, new firms, entrepreneurs and technologies come into play, along with a willingness among market actors to accept a certain level of risk to secure the benefits afforded to the risk taker under the rules of the law of capture. The promise (often illusory) of great reward provides the motivation for commercial activity that brings together all of these elements. While risks are usually great in natural resource development, one area where risk can be minimized is in the area of property rights. An important role of government is to uphold contracts and to protect the property rights of indi-viduals and corporations that assume the risks inherent in entrepreneurial enter-prises, particularly in the area of energy discovery and recovery. As the United States seeks to capitalize on the growth of both fossil and renewable energy resources on public and private lands, it must take care not to unreasonably add costs to those individuals and firms acting to maximize energy security while also seeking to make a profit. This set of required elements is needed for both renewable energy firms and fossil energy firms, as virtually all forms of energy are currently needed to address the multiple dimensions of the energy security dilemma. Simultaneously, while being sensitive to the rights of individuals and firms, public policy must address public concerns and needs. The United States has a unique property rights arrangement that allows individual or corporate entities to own sub-surface mineral rights. Royalties paid to the federal govern-ment for resources extracted from public lands are historically low, serving as a powerful incentive to discovery. For resources extracted on private lands, royalties paid to the federal government are non-existent. However, the extraction of the resource and the longstanding property right arrangements and their precedents, contribute to high marginal social costs and greater energy insecurity. Mineral rights can be privately held, yet many of the short-term and nearly all of the long-term externalities of developing these rights are borne by society at large. This situation means, in effect, that as a society we have privatized the gains and socialized the costs of such energy development. Thinking back to James Q. Wilson's policy typology, our current mineral rights arrangements exist somewhere between interest group politics and client politics. However, a more equitable long-term solution to energy security may be served by the incorporation of some degree of added focus on community sustainability and majoritarian political solutions.[5]

Extending the imagination beyond the possibilities suggested above, one can envision other methods of enhancing energy security by reducing energy demand—albeit over the long term—that involve rather dramatic changes to the way products are manufactured and transported to market. Newly developed 3-D printing technology has just begun to demonstrate its potential to allow for many types of goods to be manufactured locally. This newly emerging technological revolution has already allowed for the production and use of items as diverse as prosthetic limbs and auto parts. Other possibilities include the production of food and clean energy. One experimental effort has already yielded a small wind turbine manufactured on a 3-D printer.[6] Technological capacity of this sort could reduce dramatically the demand for fossil energies transported long distances from energy-producing nations to local markets. No longer might the United States rely on nations such as China for a myriad of manufactured products. Rather, production lines could be brought to domestic shores or even brought directly into the individual consumer's home. Reliance on foreign petroleum could decline dramatically, and the nation's trade deficit would decline as well.

In this book we have shown how energy security is comprised of many objectives, and the importance (along with the difficulty) of identifying the marginal social costs and benefits associated with their pursuit. Energy security has interconnected domestic and international dimensions, and entails many competing goals and policies which often work at cross purposes—bringing to mind the kaleidoscope and patchwork quilt metaphors introduced earlier. Moreover, its definition and pursuit are constantly evolving as conditions in technology, domestic and global markets, public policy and international relations change. As has been the case for quite some time, finding a balance among energy security aims remains an elusive goal. While the United States produces increasing levels of fossil energy per day, numerous federal, state and local government agencies work tirelessly to promote sustainability and the subsidization of clean energy for carbon-based sources. Agreement on the goals of energy security, much less the means to achieve them, is often resistant to policy initiatives, which are caught in a combination of cross-cutting interests, along with partisan, regional, and cross-generational politics.

At the same time, while energy security appears to be a looming problem that requires a large and long-lasting commitment by governmental and private institutions, it must also be understood that much of the energy security dilemma—along with its alleviation—is a function of individual behavior. Per capita energy consumption rates are rising globally. We all make choices when we consume. Likewise, we make choices when we choose to take a well-paying job downtown but live 40 or 50 miles away in a suburb, in the foothills, or out among the fields and trees of the countryside. We make choices when we fly to Europe or take a cruise. Each of those individual choices may increase our level of happiness and fulfillment, but at the time we make these individual decisions we likely think precious little about energy security. Rather, we think principally of the salary we will earn or the fun we will have on our trip. Policymakers may wring their hands, activists may lament the environmental consequences, and energy companies may embrace their good fortune, but in some sense we are reflecting our values, which generally place a very high "discount" on the future and a relatively high value on the present every time we make a purchase—and that includes flipping on a light switch in our homes. We are effectively saying that the risk is acceptable because we are paying a monetized price for both the energy and the risk involved in obtaining it. As this book has argued, however, every energy source has its costs, many of which are less visible and/or externalized, and none provides a singular path to the achievement of energy security. But there is no free lunch, and so a clearer understanding of how we pay for our lunch is necessary for a more honest assessment of those costs we are willing to pay, and those we are not.

Notes

1 US Energy Information Administration, "U.S. Remained World's Largest Producer of Petroleum and Natural Gas Hydrocarbons in 2014," *Today in Energy*, April 7, 2015.
2 Export-Import Bank of the United States, "US Exporters to Benefit from a $2 Billion Preliminary Commitment to Brazil's Petrobras," May 6, 2009, www.exim.gov/news/us-exporters-benefit-2-billion-ex-im-bank-preliminary-committment-brazils-petrobras (accessed September 17, 2015).
3 Jeremy Page, "China Presence Complicates Sea Drills: New Political and Legal Challenges Emerge with Beijing's Decision to Send Its Navy to Train Alongside Rivals in Pacific," *The Wall Street Journal*, July 17, 2014, http://online.wsj.com/news/articles/SB40001424052702303833804580018923852227094 (accessed July 19, 2014); Brian Spegele and Vu Trong Khanh, "China Moves Oil Rig From Contested Waters," *The Wall Street Journal*, July 6, 2014, http://online.wsj.com/articles/chinas-cosl-moves-oil-rig-from-contested-waters-1405472611 (accessed July 19, 2014).
4 Amory Lovins and the Rocky Mountain Institute, *Reinventing Fire: Bold Solutions for the New Energy Era*, White River Junction: Chelsea Green Publishing, 2011.
5 James Q. Wilson, *Bureaucracy: What Government Agencies Do and Why They Do It*, New York: Basic Books, 1989.
6 3DP Applications, *3Dprintingindustry.com*, http://3dprintingindustry.com/2015/02/27/3d-printed-wind-turbines-help-remote-communities-gain-sustainable-power/ (accessed May 14, 2015).

INDEX